Springer Theses

Recognizing Outstanding Ph.D. Research

Aims and Scope

The series "Springer Theses" brings together a selection of the very best Ph.D. theses from around the world and across the physical sciences. Nominated and endorsed by two recognized specialists, each published volume has been selected for its scientific excellence and the high impact of its contents for the pertinent field of research. For greater accessibility to non-specialists, the published versions include an extended introduction, as well as a foreword by the student's supervisor explaining the special relevance of the work for the field. As a whole, the series will provide a valuable resource both for newcomers to the research fields described, and for other scientists seeking detailed background information on special questions. Finally, it provides an accredited documentation of the valuable contributions made by today's younger generation of scientists.

Theses are accepted into the series by invited nomination only and must fulfill all of the following criteria

- They must be written in good English.
- The topic should fall within the confines of Chemistry, Physics, Earth Sciences, Engineering and related interdisciplinary fields such as Materials, Nanoscience, Chemical Engineering, Complex Systems and Biophysics.
- The work reported in the thesis must represent a significant scientific advance.
- If the thesis includes previously published material, permission to reproduce this must be gained from the respective copyright holder.
- They must have been examined and passed during the 12 months prior to nomination.
- Each thesis should include a foreword by the supervisor outlining the significance of its content.
- The theses should have a clearly defined structure including an introduction accessible to scientists not expert in that particular field.

More information about this series at http://www.springer.com/series/8790

Thomas Helmer

Development of a Methodology for the Evaluation of Active Safety using the Example of Preventive Pedestrian Protection

Doctoral Thesis accepted by
Technische Universität Berlin, Germany

 Springer

Author
Dr. Thomas Helmer
Fakultät V Institut für Land- und
    Seeverkehr, Fachgebiet Kraftfahrzeuge
Technische Universität Berlin
Berlin
Germany

Supervisor
Prof. Dr. Volker Schindler
Fakultät V Institut für Land- und
    Seeverkehr, Fachgebiet Kraftfahrzeuge
Technische Universität Berlin
Berlin
Germany

ISSN 2190-5053   ISSN 2190-5061   (electronic)
ISBN 978-3-319-12888-7   ISBN 978-3-319-12889-4   (eBook)
DOI 10.1007/978-3-319-12889-4

Library of Congress Control Number: 2014953923

Springer Cham Heidelberg New York Dordrecht London

Printed on acid-free paper

Springer is part of Springer Science+Business Media (www.springer.com)

Part of this thesis have been published in the following documents:

Journals and Books

Ebner, A., Helmer, T., Samaha, R. R., and Scullion, P. Identifying and Analyzing Reference Scenarios for the Development and Evaluation of Active Safety: Application to Preventive Pedestrian Safety. *International Journal of Intelligent Transportation Systems Research 9*, 3 (2011), 128–138

Helmer, T., Scullion, P., Samaha, R. R., Ebner, A., and Kates, R. Predicting the Injury Severity of Pedestrians in Frontal Vehicle Crashes based on Empirical, In-depth Accident Data. *International Journal of Intelligent Transportation Systems Research 9*, 3 (2011), 139–151

Kompass, K., Huber, W., and Helmer, T. Safety and comfort systems: Introduction and overview. In *Handbook of Intelligent Vehicles*, A. Eskandarian, Ed. Springer, 2012

International Conferences

Helmer, T., Ebner, A., and Huber, W. *Präventiver Fußgängerschutz - Anforderungen und Bewertung*. 18. Aachener Kolloquium Fahrzeug- und Motorentechnik, 2009

Ebner, A., Helmer, T., and Huber, W. Bewertung von Aktiver Sicherheit - Definitionen, Referenzsituationen und Messkriterien. In *1. Automobiltechnisches Kolloquium; München, 16. und 17. April 2009, Technische Universität München-Garching* (Düsseldorf, 2009), VDI Wissensforum GmbH

Helmer, T., Samaha, R. R., Scullion, P., Ebner, A., and Kates, R. Injury risk to specific body regions of pedestrians in frontal car crashes modeled by empirical, in-depth accident data. In *Proceedings of the 54th Stapp Car Crash Conference* (2010)

Helmer, T., Scullion, P., Samaha, R. R., Ebner, A., and Kates, R. Predicting the injury severity of pedestrians in frontal car crashes based on empirical, in-depth accident data. In *Proceedings of the 17th ITS World Congress* (2010)

Helmer, T., Samaha, R. R., Scullion, P., Ebner, A., and Kates, R. Kinematical, physiological, and vehicle-related influences on pedestrian injury severity in frontal car crashes: multivariate analysis and cross-validation. In *Proceedings of the International Research Council On Biomechanics Of Injury (IRCOBI)* (2010), pp. 181–198

Kates, R., Jung, O., Helmer, T., Ebner, A., Gruber, C., and Kompass, K. Stochastic simulation of critical traffic situations for the evaluation of preventive pedestrian protection systems. In *Erprobung und Simulation in der Fahrzeugentwicklung* (2010)

Ebner, A., Samaha, R. R., Scullion, P., and Helmer, T. Identifying and analyzing reference scenarios for the development and evaluation of preventive pedestrian safety systems. In *Proceedings of the 17th ITS World Congress* (2010)

Ebner, A., Samaha, R. R., Scullion, P., and Helmer, T. Methodology for the development and evaluation of active safety systems using reference scenarios: application to preventive pedestrian safety. In *Proceedings of the International Research Council On Biomechanics Of Injury (IRCOBI)* (2010), pp. 155–168

Helmer, T., Ebner, A., Jung, O., Paradies, S., Huesmann, A., and Praxenthaler, M. Fahrerverhalten in Fußgängersituationen mit und ohne Unterstützung eines präventiven Sicherheitssystems - Herausforderungen bei der empirischen Bewertung. In *AAET 20011 - Automatisierungssysteme, Assistenzsysteme und eingebettete Systeme für Transportmittel*. Gesamtzentrum für Verkehr Braunschweig e.V., 2011

Helmer, T., Neubauer, M., Rauscher, S., Gruber, C., Kompass, K., and Kates, R. *11th International Symposium and Exhibition on Sophisticated Car Occupant Safety Systems*. Fraunhofer-Institut für Chemische Technologie ICT, 2012, ch. Requirements and methods to ensure a representative analysis of active safety systems, pp. 6.1–6.18. ISSN 0722-4087

Helmer, T., Kühbeck, T., Gruber, C., and Kates, R. Development of an integrated test bed and virtual laboratory for safety performance prediction in active safety systems (F2012-F05-005). In *FISITA 2012 World Automotive Congress - Proceedings and Abstracts* (2012). ISBN 978-7-5640-6987-2

Patents

Kates, R., Jung, O., Helmer, T., Ebner, A., and Gruber, C. Verfahren zum Entwickeln und/oder Testen eines Fahrerassistenzsystems. German Patent DE 102011088807.1, 2013

Supervisor's Foreword

Newly developed driver assistance systems have the potential to provide a significant contribution to traffic safety. However, their efficacy regarding accident avoidance and mitigation is strongly influenced by their conception, technical capabilities, design, and subsequent integration. Additionally, their actual performance as intended in the field is influenced by a broad spectrum of traffic, environmental, and individual factors. Developers as well as other stakeholders in traffic safety urgently require objective as well as representative methods for assessing and predicting the efficacy of driver assistance systems. This requirement poses a complex problem due to the sheer number of influencing parameters. In particular, a prospective assessment of the potential performance of novel systems is needed during development.

Assessment must include quantification of the intended positive safety effects as well as of possible undesired consequences for the equipped vehicle or surrounding traffic. The evaluation process presented here utilizes virtual experiments using stochastic simulation techniques. The evidence basis for those experiments includes accident statistics, naturalistic driving observations, and driving simulator studies for definition of relevant conflicts and assignment of the appropriate statistical weighting. This evaluation process allows for an assessment of different driver assistance systems in a large variety of potentially critical driving situations. Based on those virtual experiments, the efficacy of the system in hazardous situations (involving, e.g., vehicles and pedestrians) is quantified by appropriate key characteristics.

The systematic approach of this thesis includes review of existing procedures, design of a new evaluation process, and proof of concept using vehicle-based preventive pedestrian protection in crossing scenarios as an example. To this end, a large sample of virtual traffic situations is generated, which sometimes result in accidents. For a correct virtual representation, all relevant human and technical processes are modeled including their possible failure modes. Since many influencing parameters are distributed, a quantification of efficacy needs to take their frequency of occurrence into account. This requires an interdisciplinary and evidence-based approach.

In order to quantify the safety benefits of virtual collisions in terms of changes in injury severity, models allowing a probability-based quantification of injury severity using characteristics at the time of impact are needed. The approach taken here is to develop a process adapted to the requirements of injury modeling based on empirical data. This process uses multivariate logistic regression in order to derive plausible as well as validated models. The definition of meaningful key characteristics, partly oriented and transferred from medical sciences and healthcare economics, advances the interpretation of results and supports sound decisions by different stakeholders.

Dr. Helmer's thesis represents a considerable advancement in the field of active vehicle safety assessment. The work enables a well-founded evaluation of the efficacy of active safety systems in traffic. Since the key characteristics are comparable to well-known parameters used in passive safety, a consistent assessment of overall safety in traffic is enabled, according to the paradigm of integral safety. The great potential of the set of methods, to which this thesis contributes, lies in its applicability for academia, industry, and policy makers, such as legislation and consumer protection agencies, all sharing the common denominator of promoting measures with the highest real efficacy in traffic.

Berlin, July 2014   Prof. Dr. Volker Schindler

Acknowledgments

First of all, I would like to thank all Professors who supported me during the years I have been working on this thesis. Very special thanks go to Prof. Dr. rer. nat. Volker Schindler, Leiter Fachgebiet Kraftfahrzeuge at Technische Universität Berlin, for supervising the thesis. His ideas, encouragement, and quiet guidance have been an enormous help.

This Ph.D. thesis has been conducted during my work at the Vehicle Safety Department of BMW Group in Munich. It is a pleasure for me to thank those who made this thesis possible. Representative for my supervisors in the last three years, I want to thank Mr. Klaus Kompaß, Vice President Vehicle Safety, for his support. Thanks to his confidence in me and my work; I was given the chance to participate in highly interesting projects, conduct a research stay abroad, and present the results at various occasions to the international automotive and scientific community.

I am indebted to my many colleagues who have supported me over the last years. Especially, I want to thank Mr. Adrian Ebner for many stimulating and enriching discussions. One highlight of our work was a joined research stay at the National Crash Analysis Center (NCAC) of the George Washington University (GWU). I am grateful for all the help from Randa Radwan Samaha, MS, Paul Scullion, BS, and Dr. Kennerly Digges, all of them with GWU, who made this experience possible.

I owe my deepest gratitude to Dr. Werner Huber, who gave me continuous advice as a scientific mentor as well as supervisor and close colleague. Without his trust, encouragement, and uncompromising support, this research would have been much more difficult to conduct and definitely less fun.

The contribution of Dr. Ronald E. Kates has to be mentioned especially. I feel honored to be given the chance to learn from him during many discussions and work with him closely on projects and various publications.

It is an honor for me to thank Prof. Dr.-Ing. Dr.-Ing. E.h. Hans-Hermann Braess for the time he spent in many mind-opening and stimulating discussions. The scientific input and the possibility to profit from his enormous experience and knowledge in vehicle safety meant tremendous help during the whole research.

Last but not least I want to thank my family, friends, and especially Maria for their continuous understanding, motivation, and support during the last years.

Munich, September 2013   Thomas Helmer

Contents

Author Biography

Dr. Thomas Helmer received his Diploma (M.Sc. degree) in Engineering Sciences from Technische Universität München, Munich, Germany, in 2007 with high distinction. He began his doctoral research in March 2008 within a collaborative effort between the Technische Universität Berlin, Germany, and the BMW Group in Munich. His doctoral work, under the supervision of Prof. Dr. Volker Schindler, focused on the evaluation of active safety in traffic. Thomas Helmer received his Dr.-Ing. (Ph.D. degree) from Technische Universität Berlin in 2014. During his doctoral study, he conducted research as an official visitor to the National Crash Analysis Center of The George Washington University in Washington, D.C., sponsored by a scholarship of the German Academic Exchange Service (DAAD). The findings of the thesis have been published in internationally well-recognized journals and conferences. Thomas Helmer received the Outstanding Paper Award of the 17th ITS World Congress in 2010. He holds a patent and is an inventor on several pending patent applications.

Chapter 1
Introduction

1.1 Safety in Road Traffic

Vehicle-based road traffic as well as road safety affect people in every country. Analysis of this issue requires an understanding of the fundamental relationships and effects of road traffic. To this end, this thesis starts with a short summary on basics regarding traffic participants, their interaction, and the relation to safety, concentrating on the driver and pedestrian. As this work focuses on vehicle-based approaches, additional safety measures such as education or changes to infrastructure are not explicitly considered here.

Safety (and other) characteristics arise for the driver from his interaction with the vehicle and the environment (including all other participants) [1]. The dynamical nature of these interactions in traffic can be illustrated by a control theory model [2–4]. Here, the human driver represents a complex controller, selecting the route and carrying out actions at several levels in such a way as to keep target variables (such as car following gaps) within a desired range, while responding to multiple inputs and feedback from both the environment and the vehicle.

This basic controller scheme can be extended to include driver assistance, in particular the elements of active safety systems. Conceptually, "active safety systems" comprise all measures contributing to avoidance of accidents or mitigation of their severity, prior to the collision [5, 6]. Of course, they are part of the vehicle, but they differ from "standard" vehicle controls by actively interacting with the driver, with the standard vehicle controls, and with the environment [7–9]. These systems compile information about the vehicle, the environment, and the driver; assess and interpret this information, using internal system models, and calculate a target behavior or response. If the current state deviates from this target, a driver assistance system calculates the appropriate action or feedback. Possible system responses can include information or warnings to the driver or other participants as well as automatic interventions in vehicle dynamics. The intensity of the response of the system depends on the design characteristics of the particular system, the reliability of algorithms for interpreting and classifying the current driving state and traffic situation, the inferred

© Springer International Publishing Switzerland 2015
T. Helmer, *Development of a Methodology for the Evaluation of Active Safety using the Example of Preventive Pedestrian Protection*, Springer Theses,
DOI 10.1007/978-3-319-12889-4_1

criticality of the situation and the degree to which a driver response "in the loop" could be expected within the time available.

Effective active safety systems usually require a well-coordinated interaction of all elements of the control loop. The primary driving task can be supported at any level of the following three-level hierarchy [1]:

- The high-level *navigation* task is derived from the desire to achieve a particular objective of a trip and to reach a specific destination; it comprises route planning and estimation of travel time, including possible adaptations of the route to changing traffic or other boundary conditions.
- At the intermediate level, the *driving control* task requires monitoring and adjustment of target variables, such as choosing a lane and a desired speed, while taking into account boundary conditions and external influences such as the dynamics of the traffic flow. Driving maneuvers are carried out in order to fulfill the navigation task.
- The lowest level of the model represents the process of *stabilization*; it includes all tasks that keep the vehicle "on course" (e.g., steering and braking).

The driving state is normally continually monitored (by the driver and/or a system) in order to make corrections on any or all of these levels if required. Detailed applications, variations and refinements of this model can be found in the literature [4, 10–12]. Classically, active safety systems, e.g., Dynamic Stability Control (DSC), have been designed to provide support at the stabilization level. At this level, the target quantities are generally well defined in terms of vehicle physics. Preventive pedestrian protection, which is in the focus of this thesis, addresses primarily the maneuvering level and thus involves additional complexities in control—particularly those involving the interpretation of driver behavior and the interaction of system actions with the driver.

Since the driver's role in the control loop is decisive, it is helpful to consider the characteristics of the driver and his behavior in detail. A classical hierarchical behavior model for target-oriented actions has been described by Rasmussen [12]. This model distinguishes three categories of "cognitive demands on humans in work processes": knowledge-based, rule-based, and skill-based behavior. If a person is confronted with complex tasks requiring untrained actions or reactions, the cognitive demands result in "knowledge-based" behavior. In this behavioral mode, possible actions are first mentally reviewed before the strategy that appears to provide the best solution is implemented. People will generally carry out "rule-based" behavior in situations that they have repeatedly experienced, drawing on an inventory of learned rules or behavior patterns. These readily available rules and patterns allow a faster response to the situation. "Skill-based" behavior arises whenever situational demands have been trained in a learning process and stimulus-response mechanisms are characterized by reflexive actions. Responses and performance are fastest at this level due to the routine and essentially autonomous execution of processes and actions.

By taking the driving task requirements and the behavioral level into account, one can compare the time required by the driver with the time available to him for particular situations and maneuvers. This comparison facilitates an estimation of the driver's needs and the potential benefits of an assistance system [13]. Coping with (highly) critical pedestrian situations could involve knowledge-based behavior, due to the rare occurrence of these interactions. Another kind of possible reaction could be an unconscious reflex, thus being fast but probably inadequate. As a consequence, finding and carrying out the optimal reaction for handling the situation will usually require more time than is available. Hence, failures will occur more frequently in comparison to other situations with a similar time budget but that can be addressed at a lower (e.g., skill-based) behavioral level.

Following Reichart [13] a generic fault tree can be used in order to illustrate the logical relationships and causal structure of events leading up to an accident (see Fig. 1.1). The analysis begins at the stage where the participants (i.e., the driver and/or pedestrian) make particular mistakes and get involved in a traffic conflict. A traffic conflict may be characterized by considering approaching object trajectories which, extrapolated in time, would exhibit an increased probability for collision unless one of the participants changes his current state of motion [18]. The influences of human performance shaping factors and resulting mistakes leading to a conflict can be found in the literature [19–22]. If the conflict cannot be resolved (even by intervention of a preventive system), an accident will occur. An accident in road traffic is defined as an event that occurs suddenly, is connected to the typical dangers in road traffic, and results in personal injury or significant property damage [23].

Projecting the fault tree model onto a time line results in the phases of an accident (see Fig. 1.2). From "normal" driving up to a traffic conflict, a system may provide warnings or may intervene in vehicle dynamics directly. As the situation becomes increasingly critical, a "point of no return" may be reached, where a collision is physically inevitable. By definition, active safety is designed to be effective before physical contact. From the moment of physical contact on, the event is considered as an accident. During the crash phase, passive safety design is essential for reducing

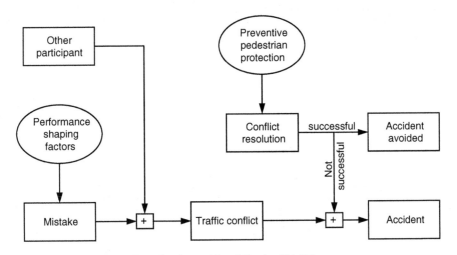

Fig. 1.1 Generic genesis of a pedestrian accident following [14–17]

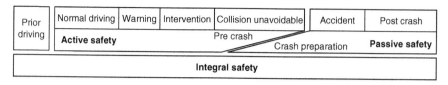

Fig. 1.2 Phases of an accident (following [27])

Table 1.1 The Haddon Matrix (following [26])

	Pre crash Crash prevention	Crash Injury prevention	Post crash Life sustaining
Human	Information Attitudes Impairment Police enforcement	Use of restraints Impairment	First-aid skill Access to medics
Vehicles and equipment	Roadworthiness Lighting Braking	Occupant restraints Other safety devices Crash protective design	Ease of access Fire risk Automatic crash notification
Environment	Road design and road layout Pedestrian infrastructure Speed management	Crash protective roadside objects	Rescue resources Congestion

accident consequences [24]. The post-crash phase is characterized by rescue of the injured. The three phases of an accident combined with the model of driver, vehicle, and environment together with possible countermeasures are summarized in the well-known Haddon-Matrix [25, 26] (Table 1.1).

The concept of Integral Safety is a holistic approach to vehicle safety including active and passive safety as well as the direct interaction of both. The primary objective is to provide a high level of safety during all phases of an accident as a result of an effective combination and interaction of measures of both active and passive safety [14, 17, 27, 28].

1.2 Accident Statistics

The magnitude of the "social illness" traffic safety is best displayed when looking at the global dimension. It is estimated that more than about one million people die each year in road traffic (about 1.24 million in 2010) and that between 20 and 50 million people suffer injuries [29]. (The range of injury estimates reflects differences in reporting systems and schemes between countries.) The total number of accidents is difficult to estimate [29]. Road traffic injury ranks number eight on the list of leading causes for death and is expected to reach number five in 2030, so that its importance is increasing [29]. Aside from personal and social implications, traffic accidents and injuries also have a considerable economic impact. Global losses are estimated to be US$518 billion and can amount to between 1 % and 3 % of the gross national product per year [30].

The phases of an accident can also be analyzed by frequency of occurrence in road traffic. Figure 1.3 displays the corresponding proportions as a pyramid, in which the vertical axis represents the severity of the event and the volume of each segment stands for its frequency in road traffic. It is assumed (Hydén 1997 quoted by [18]), that there are close causal connections between conflicts and accidents in traffic, but the actual transition probability is hard to quantify [18]. The only representative quantification available refers to accidents. The percentages in Fig. 1.3 quantify the "volume" of the segments for the year 2010 in Germany and the US. Accidents are the "tip of the pyramid". Most of the accidents involve property damage only; in a small proportion, there are injured persons, and a very few lead to fatal injuries.

In order to highlight some key safety issues, it is useful to explain the trends in Germany and the US. Accidents in traffic are statistically rare events [33, 34]. The accident incidence based on km driven as a measure of exposure was on average once every 292,000 km (Germany) and accordingly 881,000 km (US) in 2010. Pedestrian accidents occur even less frequently: in Germany, about once every 23,000,000 km and in the US once every 68,000,000 km. [Note that these ratios are intended only to illustrate the rarity of the event of an accident and are not appropriate for characterizing the overall level of traffic safety in either of the countries mentioned. Depending

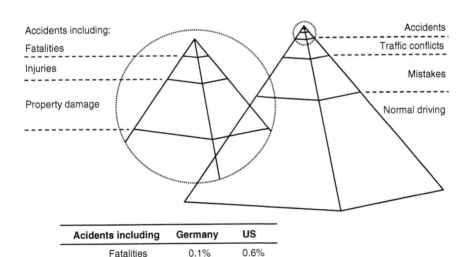

Acidents including	Germany	US
Fatalities	0.1%	0.6%
Injuries	11.9%	27.9%
Property damage	88.0%	71.5%

Fig. 1.3 Levels of interaction in road traffic including quantification of accidents for Germany and the US in 2010 [31, 32]

Table 1.2 Accident ratios for Germany and the US in 2010

Exposure	Injured persons		Fatalities	
	Germany	US	Germany	US
1,000 accidents	154	413	1.51	6.07
1,000 accidents with injured	1,300	1,452	12.65	28.92
10^9 km traveled	532	469	5.17	6.89
100,000 population	458	724	4.46	10.63

Originally missing corresponding ratios have been computed [31, 32]

on the research in question, different ratios are published by the agencies as indicators for traffic safety of a country (see Table 1.2).]

Although the above statistics suggest that the probability per km driven to have an accident in Germany is about 3.0 times higher than in the US, more facts are generally required for an overview of traffic safety in a particular country. The numbers in Table 1.2 describe different conditional probabilities related to injuries and their severity. The first two rows give information on the level of passive safety as well as the average intensity of the accidents (the contribution of those effects cannot be clearly separated based on overall accident statistics). Statistically speaking, this is the conditional probability of a person's being injured or killed, once they have an accident. On average, 154 people are injured per 1,000 accidents in Germany and 413 in the US (fatalities: 1.51 vs. 6.07). Considering only accidents with injuries (i.e., taking out all property damage only cases), the difference between the two countries is reduced. The statistical risk to life while participating in road traffic is expressed in row number three: It gives the probability of being injured or killed per one billion km driven. The risk of injury is higher in Germany (532 vs. 469 injured persons in the US) whereas the fatality rate is lower (5.17 vs. 6.89). The last row is commonly used, but is somewhat harder to interpret, as there is no straightforward causal connection between the size of the population or the number of licensed drivers and the probability to suffer harm in a road traffic accident. Many more factors, such as km driven, motorization, belt use, size and grade of industrialization, or geopolitical situation, influence these probabilities in general.

The economic impact in 2009 accounted for €30.5 billion in Germany [35] and for US$230.6 billion in the US [36]. Nevertheless, the official statistics for both Germany and the US show an improvement in traffic safety for the last 20 years. Figures 1.4 and 1.5 present the development in both countries including absolute values for the most recent year. Several facts become obvious from the indicators given. In both countries, the km driven increased about 45 % and 40 %, respectively. The number of accidents remained nearly constant in Germany and seems to have declined in the US. The number of injured persons decreased by about 30 % in both countries. In Germany there has been a steady decline leading to 70 % fewer fatalities in 2010 than in 1990. Fatalities (as an absolute number) have declined by about 25 % in the US, but the reduction has been quite pronounced only in the last few years. The fatality rate (fatalities per km driven) has decreased steadily in both countries.

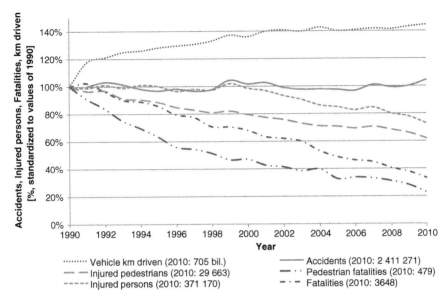

Fig. 1.4 Changes in traffic safety in Germany from 1990 to 2010, standardized for 1990 [32]

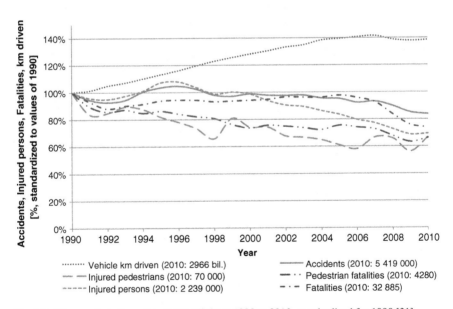

Fig. 1.5 Changes in traffic safety in the US from 1990 to 2010, standardized for 1990 [31]

Interpreting these statistics regarding the effect of active and passive safety, the following observations can be made: The decreasing accident rate (fairly constant number of accidents for increasing km driven) strongly suggests that measures for avoiding accidents have been successful, including both infrastructure measures and active safety in vehicles. On the other hand, the ratio of fatalities to accidents is a strong indicator for passive safety (including emergency services), since it reflects the conditional probability of being fatally injured given being involved in an accident. Improvements in passive safety are evident in the ratio for Germany; for the US, improvements in vehicle-based passive safety may have been masked by additional factors such as speed limit increases, a tendency for unbuckled motorists, etc.

Figures 1.4 and 1.5 also present the corresponding trends for pedestrians. In both Germany and the US, accidents with pedestrians as well as the number of injured pedestrians have decreased more rapidly than total accidents or injuries as a whole. However, the rate of improvement regarding pedestrian fatalities seems to have slowed down in the last decade. (Note that the curves for total pedestrian accidents and injured pedestrians are virtually parallel, since normally a pedestrian accident involves mostly one injured person, the pedestrian.) The numbers for fatalities differ in Germany and the US; both show a significant reduction, about 75 % and 35 % respectively.

Although there have been advances in traffic safety in Germany and in the US, there is still a need for action regarding traffic safety in general and the protection of pedestrians especially. The issue is drawing increasing attention in the legislative, scientific and industrial community, as well as in consumer protection groups. In terms of absolute numbers (given in Figs. 1.4 and 1.5) only 1.3 % (Germany) and accordingly 1.3 % (US) of all accidents did involve pedestrians in 2010. However, considering injury severity, 8.0 % of all injured persons in Germany were pedestrians (respectively 3.1 % in the US). The vulnerability of pedestrians can be most clearly seen by considering fatalities: 13.1 % of all fatalities are pedestrians in Germany and 13.0 % in the US. Defining the protection of humans as the highest priority in traffic safety and considering recent trends in accident statistics, protecting pedestrians thus emerges as top priority considering their disproportionately high fraction among injuries and fatalities.

1.3 Pedestrian Protection

For the development and enhancement of measures of active safety, a top-down or goal-directed approach offers advantages compared to a technology-driven, bottom-up development of functions. At the top level (Fig. 1.6), the goal is safety as a characteristic of a vehicle. A basic scheme distinguishes the following levels: vehicle characteristics (e.g., safety, comfort, design), functions, systems, and components, providing a structure linking characteristics to realizations. A function is defined as a solution-neutral requirement for the realization of a characteristic. A system and its components represent the actual technical realization of the function in question.

Figure 1.6 illustrates the structure of active safety as a characteristic. One function of active safety is "avoidance of collisions". One of the systems able to fulfill this

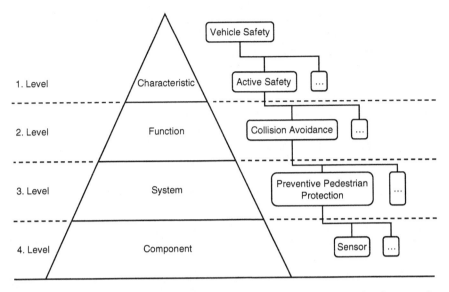

Fig. 1.6 Top-down structuring of active safety with an example for preventive pedestrian protection

function might be preventive pedestrian protection with its specific components. It is vital for a successful and effective fulfillment of a characteristic (i.e., having a high efficacy in active safety) to structure the problem and derive solutions from the top down instead of just starting with a particular technical solution. In this way, the search for a technical solution optimizing the required characteristic can be begun at the functional level. This procedure facilitates a systematic search for system and component solutions and a comparative evaluation of possible technical alternatives.

The evaluation of active safety as a characteristic is not carried out at the level of components or systems, but focuses on the characteristic itself or the function. The two lowest levels are evaluated during system development. With this scheme in mind, a short summary of existing technology and the corresponding laws, regulations, and consumer protection initiatives focusing on pedestrians is given in the following.

The minimum requirements for vehicle-based pedestrian protection necessary for vehicle type approval ("homologation") are defined by laws, e.g., the Regulation (EC) No. 78/2009 of the European Parliament and of the Council [37] or the Global technical regulation No. 9 on Pedestrian Safety by the United Nations [38]. Additional requirements are defined by consumer protection agencies like the European New Car Assessment Program (Euro NCAP) [39].

At present, pedestrian protection requirements mainly involve optimization and implementation of passive safety measures at vehicle front ends, where most pedestrian impacts occur. The most prominent improvements have involved changes in the shape of the front end and the elimination of sharp or rigid mounted parts, e.g., bull bars [40], in order to minimize obvious sources of injury. Another important passive measure is the provision of adequate space for deformation, mainly in the hood and bumper areas, with the intention of allowing the absorption of increased proportions of kinetic energy [41, 42].

Design changes in response to the requirements of passive pedestrian protection can interact with other design objectives. In order to further improve safety performance and eliminate or reduce potential conflicts between safety and design objectives [43], research and development have intensified work on novel, advanced approaches to passive protection. Typical representatives of such systems are the pop-up hood and the pedestrian airbag. Both of these are in fact "active" devices. However, they are still considered as passive safety features because of their functionality, i.e., to provide more "soft" deformation space before impact on hard components. The pop-up hood deploys a few moments before the impact of the pedestrian on the hood. It allows for additional absorption of energy, which is especially important for the head as the most critical body part determining vehicle-pedestrian accident severity [44, 45]. Whereas this component is already installed in new cars [46], the pedestrian airbag is just making the transition from R&D to implementation. The basic idea is that an airbag on the outside of the vehicle provides the deformation space in front of classically rigid components, such as the A-pillar or the lower end of the windshield [47].

All passive safety measures implemented in the vehicle are only capable of addressing the so-called "primary collision" (i.e., contact with the vehicle). For example, only about 6 % of pedestrians impact with the head on the hood, which limits the efficacy of measures implemented there [48]. Secondary collisions, i.e., contacts with the road surface or other objects, are not addressed by those measures. In contrast active safety measures address the entire sequence of events and as a consequence have a much higher injury avoidance potential [17, 49]. Preventive systems are at the moment in the state of development or already in the market [48, 50–52].

The first consideration of active safety in regulations is included in (EC) No. 78/2009 [37], Chapter III Article 11. All "vehicles equipped with collision avoidance systems may not have to fulfill the test requirements laid down in Sections 2 and 3 of Annex I in order to be granted an EC type-approval or a national type-approval for a type of a vehicle with regard to pedestrian protection, or to be sold, registered or to enter into service". It is required that "[a]ny measures proposed shall ensure levels of protection which are at least equivalent, in terms of actual effectiveness, to those provided by Sections 2 and 3 of Annex I". Article 11 provides a legal basis for future fulfillment of the regulation by both active and passive safety devices, based on the effectiveness required.

The approval process for braking assistance can be regarded as a prerequisite for active safety systems. The requirements formulated by the Directive 2003/102/EG, Phase 2, of the European Parliament and of the Council [53] could nearly not be fulfilled by means of passive safety. As a consequence, an evaluation regarding the effectiveness of different measures of pedestrian protection has been carried out [54, 55]. The commitment of the European Automobile Manufacturers' Association (ACEA) European Automobile Manufacturers' Association to implement brake assist, an active safety system, in every new car, led to a reduction of the requirements as stated in Phase 2 [56, 57]. Following those proceedings, a new approach was made by the European Union [58], which led to (EC) No. 78/2009. This development can be interpreted as a first step towards consideration and assessment of safety as a characteristic following the idea of integral safety (see Fig. 1.6) instead of a regulation of passive *or* active safety (or more precisely: specific components).

1.4 Objective and Methodological Approach

In view of the accident statistics mentioned above, the social and economic implications, and the impact on the individuals concerned, vehicle-based protection of vulnerable road users, especially pedestrians, represents perhaps the single most important challenge in vehicle safety. Recognizing the importance of this problem, the regulatory perspective has intensified demands and incentives for further improvement in pedestrian safety. However, considering the variety of conceivable solutions, each with its corresponding research, development, and testing cycles as well as technical requirements, it is crucial to identify at an early stage those approaches and technologies that are both technologically feasible and maximally effective. Technology-independent effectiveness evaluation is a key challenge for regulatory agencies, too.

Testing procedures and evaluation schemes for passive safety are defined and standardized in the regulations cited above and have reached a rather advanced stage of development. However, objective, reliable, representative and reproducible methods for evaluating the effectiveness of active safety systems, especially preventive pedestrian systems, have yet to be developed.

The objective of this thesis is the advancement of knowledge in order to enable the development of a method for evaluating active safety systems. The example used is vehicle-based preventive pedestrian protection.

This chapter gives the context of this thesis by describing the participants and interactions in road traffic and the genesis of an accident. Accident statistics, safety development process, regulations, and current technical solutions highlight the objective of this thesis.

The current state of scientific and technical knowledge regarding evaluation of active safety is summarized in Chap. 2, and the need for an evaluation method is described. To this end, different means of evaluation methods using controlled experiments, naturalistic driving data, as well as approaches based on accident data are discussed. Testing procedures and evaluation metrics in current use and under investigation are introduced.

A new approach to evaluation of active safety is then developed in Chap. 3. The process, including information needed, is described, and the prerequisites are defined. In detail, accident scenarios, configuration of a functional demonstrator of a preventive pedestrian protection system, and the simulative technique required are described. An introduction of a metric for the quantification of the change in safety rounds up the method.

Chapter 4 gives methodological research on driver behavior in response to a preventive protection system. The acceptance of false system actions in particular is investigated using driver interviews and ratings. A special objective of this experiment was to test whether a highly critical situation, which should lead to an accident under baseline conditions, can be reproducibly investigated using realistic parameters for the scenario by means of a dynamic driving simulator.

The development of injury probability models using empirical accident data is given in Chap. 5. The statistical methods used and the models obtained are discussed in detail. In order to use several models for different outcome levels at once, plausibility criteria are defined. To ensure this plausibility, a method using conditional probability identity is developed and discussed. This method as well as the possible challenge while implementing models for different outcome categories at once is not yet documented and solved in the literature. For both cases, assessment of one or several outcome categories, the methodology as well as the fully developed models are given for the Injury Severity Score (ISS) and fatalities as outcome categories.

Chapter 6 illustrates the results of the described evaluation method using different configurations of a preventive pedestrian protection system. The current results, the validity of the methodology used as well as need for further research are described. The results of parameter variations for a preventive pedestrian protection system are given using a functional demonstrator. The findings are interpreted with respect to the methodology. Metrics and processes necessary for system optimization and evaluation are introduced and discussed.

A discussion of the approaches and results and a conclusion form the last part of this thesis.

References

1. Bernotat, R. (1970). Anthropotechnik in der Fahrzeugführung. *Ergonomics, 13*(3), 353–377.
2. Bubb, H. (1993). Informationswandel durch das System. In H. Schmidtke (Ed.), *Ergonomie.* München: Carl Hanser Verlag.
3. Donges, E. (1978). Ein regelungstechnisches Zwei-Ebenen-Modell des menschlichen Lenkverhaltens im Kraftfahrzeug. *Zeitschrift für Verkehrssicherheit, 24*(2), 98–112.
4. Ehmanns, D., Gelau, C., Nicklisch, F., & Wallentowitz, H. (2000). *Zukünftige Entwicklung von Fahrerassistenzsystemen und Methoden zu deren Bewertung.* 9. Aachener Kolloquium Fahrzeug- und Motorentechnik.
5. Dietsche, K.-H. (Ed.) (2007). *Kraftfahrtechnisches Taschenbuch* (No. 26). Wiesbaden: Vieweg Verlag.
6. Huber, W., Steinle, J., & Marquardt, M. (2008). Der Fahrer steht im Mittelpunkt - Fahrerassistenz und Aktive Sicherheit bei der BMW Group. In *24. VDI / VW-Gemeinschaftstagung - Integrierte Sicherheit und Fahrerassistenzsysteme.* VDI-Berichte (Vol. 2048, pp. 123–137). Düsseldorf: VDI Verlag.
7. Bubb, H. (2001). Haptik im Kraftfahrzeug. In T. Jürgensohn & K. P. Timpe (Eds.), *Kraftfahrzeugführung.* Berlin: Springer.
8. Knapp, A., Neumann, M., Brockmann, M., Walz, R., & Winkle, T. (2009). *Code of Practice for the Design and Evaluation of ADAS.* No. V 5.0.
9. Naab, K. (2000). Automatisierung im Straßenverkehr. In *Automatisierungstechnik.* München: Oldenbourg Wissenschaftsverlag GmbH.

10. Bubb, H. (2003). Fahrerassistenz - primär ein Beitrag zu Komfort oder für die Sicherheit? In *Der Fahrer im 21. Jahrhundert*. VDI-Berichte (Vol. 1768) Düsseldorf: VDI Verlag.

11. Kompass, K., Reichart, G. (2006) Freude am Fahren zwischen Selbstbestimmung und autonomer Technik. In *AAET 2006 - Automatisierungssysteme, Assistenzsysteme und einge-bettete Systeme für Transportmittel*. GZVB Braunschweig.

12. Rasmussen, J. (1983). Skills, rules, and knowledge; signals, signs, and symbols, and other distinctions in human performance models. *Pullman: IEEE Transactions on Systems, Man, Cybernetics, 13*, 3.

13. Reichart, G. (2001). *Menschliche Zuverlässigkeit beim Führen von Kraftfahrzeugen*. No. 7 in 22 Mensch-Maschine-Systeme. Düsseldorf: VDI Verlag.

14. Domsch, C., & Huber, W. (2008). Integrale Sicherheit - ein ganzheitlicher Ansatz für die Fahrzeugsicherheit. *17. Aachener Kolloquium Fahrzeug- und Motorentechnik*.

15. Donges, E. (1999). A conceptual framework for active safety in road traffic. *Vehicle System Dynamics, 32*, 113–128.

16. Fricke, N., Glaser, C., & de Filippis, M. (2006). Passive und Aktive Sicherheitsmaßnahmen im Kraftfahrzeug. *MMI-Interaktiv, 10*(10), 39–47.

17. Kompass, K., & Huber, W. (2009). *Integrale Sicherheit – Effektive Wertsteigerung in der Fahrzeugsicherheit*. VDA, Technischer Kongress.

18. Gstalter, H. (1983). Der Verkehrskonflikt als Kenngröße zur Beurteilung von Verkehrsabläufen und Verkehrsanlagen. Dissertation, Technische Universität Carolo-Wilhelmina Braunschweig.

19. Gründl, M. (2005). Fehler und Fehlverhalten als Ursache von Verkehrsunfällen und Konse-quenzen für das Unfallvermeidungspotential und die Gestaltung von Fahrerassistenzsystemen. Dissertation, Universität Regensburg.

20. Hacker, W. (1998). *Allgemeine Arbeitspsychologie*. Bern: Huber.

21. Theis, I. (2002). Das Steer-by-Wire System im Kraftfahrzeug - Analyse der menschlichen Zuverlässigkeit. Dissertation, Technische Universität München.

22. Verein Deutscher Ingenieure. (2002). VDI 4006 Blatt 1 - Menschliche Zuverlässigkeit Ergonomische Forderungen und Methoden der Bewertung. VDI-Richtlinie ICS 01.040.03; 03.040; 13.180; 21.020, Verein Deutscher Ingenieure.

23. Verkehrsmitteilung des Bundesgerichtshof. (2002). *BGH VerkMitt 2002 Nr. 15, VRS 59, 158*.

24. Kramer, F. (Ed.). (2006). *Passive Sicherheit von Kraftfahrzeugen*. Wiesbaden: Friedr. Vieweg & Sohn Verlag.

25. Haddon, W. (1968). The changing approach to the epidemiology, prevention, and amelioration of trauma: The transition to approaches etiologically rather than descriptively based. *American Journal of Public Health, 58*(8), 1431–1438.

26. Peden, M., Scurfield, R., Sleet, D., Mohan, D., Hyder, A. A., Jarawan, E., & Mathers, C., (Eds.). (2004). *World report on road traffic injury prevention*. Geneva: World Health Organization.

27. Braess, H.-H. (1996). Aktive und passive Sicherheit im Straßenverkehr. *Zeitschrift für Verkehrssicherheit, 42*(2), 50–52.

28. Bendak, M., (Ed.) (2010). *360 Grad - Fakten zur Nachhaltigkeit 2010*. Stuttgart: Daimler AG.

29. Toroyan, T. (Ed.). (2013). *Global status report on road safety 2013: Supporting a decade of action*. Geneva: World Health Organization.

30. Toroyan, T. (Ed.). (2009). *Global status report on road safety: Time for action*. Geneva: World Health Organization.

31. National Highway Traffic Safety Administration (NHTSA). (2012). *Traffic safety facts 2010— a compilation of motor vehicle crash data from the fatality analysis reporting system and the general estimates system*. Report DOT HS 811 659, National Highway Traffic Safety Administration, National Center for Statistics and Analysis, U.S. Department of Transportation.

32. Statistisches Bundesamt. (2011). *Verkehr - Verkehrsunfälle - 2010*. No. Fachserie 8 Reihe 7. Wiesbaden: Statistisches Bundesamt.

33. Breuer, J. (2002). Objektive Bewertung der Aktiven Sicherheit von Fahrzeugen: Kritische Betrachtung. In V. D. A. Technischer (Ed.), *Kongreß 2002 - Sicherheit durch Elektronik* (pp. 105–113). Frankfurt/Main: VDA - Verband der Automobilindustrie.

34. Reichart, G. (2002). Vom Fehler zum Unfall - Ein neuer Ansatz in der Unfallforschung. In *VDA Technischer Kongreß Sicherheit durch Elektronik (2002)* (pp. 59–76). VDA - Verband der Automobilindustrie, Frankfurt / Main: VDA Verband der Automobilindustrie.
35. Straube, M. (2011). Volkswirtschaftliche Kosten durch Straßenverkehrsunfälle in Deutschland 2009. Forschung kompakt 04/11, Bundesanstalt für Straßenwesen.
36. National Highway Traffic Safety Administration (NHTSA). (2009). *Traffic safety facts 2009 - a compilation of motor vehicle crash data from the fatality analysis reporting system and the general estimates system.* Report DOT HS 811 402, National Highway Traffic Safety Administration, National Center for Statistics and Analysis, U.S. Department of Transportation.
37. Europäische Union. (2009). *Verordnung (EG) Nr. 78/2009.* Amtsblatt der Europäischen Union.
38. United Nations. (2009). *Global Technical Rule No. 9—Pedestrian Safety.* No. ECE/TRANS/ 180/Add.9. United Nations.
39. Euro NCAP. (2010). *European New Car Assessment Programme (Euro NCAP) Pedestrian Testing Protocol.* No. V 5.1. http://www.euroncap.com.
40. Anderson, R. W. G., van den Berg, A. L., Ponte, G., Streeter, L. D., & McLean, A. J. (2006). *Performance of bull bars in pedestrian impact tests.* Report CASR020, Centre for Automotive Safety Research—The University of Adelaide.
41. Bachem, K. H. (2005). Schutzpotential von realisierbaren Lösungen zum fahrzeugseitigen Fußgängerschutz. Dissertation, Rheinisch-Westfälische Technische Hochschule Aachen.
42. Davies, H., Holford, K., Assoune, A., Trioulier, B., & Courtney, B. (2009). Pedestrian protection using a shock absorbing liquid (SALi) based bumper system. In *21st International Technical Conference on the Enhanced Safety of Vehicles (ESV 2009)* (No. 09–0027).
43. Helmer, T., Ebner, A., & Huber, W. (2009). *Präventiver Fußgängerschutz - Anforderungen und Bewertung.* 18. Aachener Kolloquium Fahrzeug- und Motorentechnik.
44. Friesen, F. A. (2003). Motorhaubenaufstellung für den Fußgängerschutz. Dissertation, Rheinisch-Westfälische Technische Hochschule Aachen.
45. Inomata, Y., Iwai, N., Maeda, Y., Kobayashi, S., Okuyama, H., & Takahashi, N. (2009) Development of the pop-up engine hood for pedestrian head protection. In *21st International Technical Conference on the Enhanced Safety of Vehicles (ESV 2009)* (No. 09–0067).
46. Preisliste E-Klasse Limousine. (2009). Stuttgart: Daimler AG.
47. Bovenkerk, J., Sahr, C., Zander, O., & Kalliske, I. (2009). New modular assessment methods for pedestrian protection in the event of head impacts in the windscreen area. In *21st International Technical Conference on the Enhanced Safety of Vehicles (ESV 2009)*, (No. 09–0159).
48. Wisselmann, D., Gresser, K., Hopstock, M., & Huber, W. (2009). Präventiver statt passiver Fußgängerschutz. In *AAET 2009 - Automatisierungssysteme, Assistenzsysteme und eingebettete Systeme für Transportmittel* (pp. 60–76). Gesamtzentrum für Verkehr Braunschweig e.V.
49. Ebner, A., Helmer, T., & Huber, W. (2009). Bewertung von Aktiver Sicherheit - Definitionen, Referenzsituationen und Messkriterien. In *1. Automobiltechnisches Kolloquium; München*, 16 und 17 April 2009. Düsseldorf: Technische Universität München-Garching, VDI Wissensforum GmbH.
50. Rasshofer, R., Schwarz, D., Biebl, E., Morhart, C., Scherf, O., Zecha, S., Grünert, R., & Frühauf, H. (2007). Pedestrian protection systems using cooperative sensor technology. In *Proceedings of the 11th International Forum on Advanced Microsystems for Automotive Applications (AMAA '07)* (pp. 135–145).
51. Rosen, E., Källhammer, J.-E., Eriksson, D., Nentwich, M., Frederiksson, R., & Smith, K. (2009). Pedestrian injury mitigation by autonomous braking. In *21st International Technical Conference on the Enhanced Safety of Vehicles (ESV 2009)* (No. 09–0132).
52. Volvo. (2010) Kollisionswarnsystem mit Fußgängererkennung und automatischer Notbremsung. http://www.volvocars.com/at/all-cars/volvo-s60/pages/pedestrian-detection.aspx, as of September 9, 2010.
53. Europäische Union. (2003). *Richtlinie 2003/102/EG.* Amtsblatt der Europäischen Union.
54. Hannawald, L., & Kauer, F. (2004). *Equal effectiveness study on pedestrian protection.* Dresden: Technische Universität Dresden.

55. Lawrence, G. J. L., Hardy, B. J., Carroll, J. A., Donaldson, W. M. S., Visvikis, C., & Peel, D. A. (2006). *A study on the feasibility of measures relating to the protection of pedestrians and other vulnerable road users.* No. UPR/VE/045/06. TRL Limited.
56. Association of European Automobile Manufacturers (ACEA). (2001). *ACEA commitment relating to the protection of pedestrians and cyclist agreement.*
57. Kommission der Europäischen Gemeinschaften. (2001). *Mitteilung der Kommission an den Rat und das Europäische Parlament - Fußgängerschutz: Selbstverpflichtung der europäischen Automobilindustrie.* No. KOM(2001) 389.
58. Kommission der Europäischen Gemeinschaften. (2007). *Vorschlag für eine Verordnung des Europäischen Parlaments und des Rates über den Schutz von Fußgängern und anderen ungeschützten Verkehrsteilnehmern.* No. KOM(2007) 560.

Chapter 2
State of Scientific and Technical Knowledge on Pre-crash Evaluation

2.1 Methodological Aspects of Evaluation

The recent development and market introduction of various active safety functions within the context of integral safety have generated a demand for evaluation methods (see Chap. 1). The key research question for the evaluation of integral safety, using the paradigm of Fig. 1.6, is:

How well does a given function perform regarding safety during the pre-crash phase at the characteristic or functional level?

As this question is often asked in this manner, two vital aspects are not explicitly named or are missing. The reference situation (i.e., the baseline of comparison) for the question as well as the validity of the expected answer (which directly depends on the method used) must be included in the question. No generally accepted definition of "safety benefit", as stated in the question, exists (neither for its meaning nor for the metrics). Examples of possible interpretations are:

- Performance of a given component in a specific test or a variety of tests.
- Performance of a measure regarding a particular accident constellation.
- Performance of a measure regarding specific injuries.

Considering the introduction and discussion about new measures of integral and active safety as a background, the safety performance generally refers to the benefit in the field. The "field" is a commonly used term for the traffic system as a whole (in markets and countries where the measure will be available). As a consequence, all possible safety effects (both positive and negative) within the traffic system as a whole have to be evaluated. The answer to the question should include a trade-off between both kinds of effects rather than the magnitude of positive effects alone.

However, for practical reasons, evaluations are often limited to accidents as reference groups (instead of traffic in general). Possible negative effects, such as aspects of controllability are excluded from analysis and only possible positive effects are

© Springer International Publishing Switzerland 2015
T. Helmer, *Development of a Methodology for the Evaluation of Active Safety using the Example of Preventive Pedestrian Protection*, Springer Theses,
DOI 10.1007/978-3-319-12889-4_2

assessed. The problem with many measures and studies cited below is that the questions as well as limitations are not stated precisely enough.

In order to answer (parts of) the question stated above, a variety of methods and procedures have been developed and discussed in recent years. This chapter summarizes the most important ones together with their advantages and challenges (as far as they are generally known).

Two aspects are of special importance in this context:

1. The validity of the method with respect to the research questions it is intended to answer.
2. How the method deals with uncertainty.

Although the first point seems to be obvious when setting up or choosing a method, it is of vital importance when drawing conclusions. In the discussion that follows, "validity" refers to comparison of the results, e.g., a method or process model, with observed empirical data; therefore validation does not directly confirm that every detail is correct. "Verification" refers to the confirmation of the correctness of each individual detail, e.g., in a laboratory setting. Generally speaking, it is only possible to verify some parts of a method in detail, for example, models of reaction sequences, which have been studied quite thoroughly in the literature [1]. Other processes such as pedestrians' reaction in acute situations are understandably difficult to verify in a laboratory setting. However, in order to improve the confidence in, e.g., modeling details, one performs validation of a large spectrum of statistics which can be measured; as the number of validated relationships increases, the sensitivity of the validation procedure to possible modifications of the detailed microscopic models increases.

The second point is more subtle: Every evaluation method either uses data sources, contains modeling to some extent, relies on assumptions, or draws conclusions using some kind of extrapolation. Each of them is subject to various kinds of uncertainty (e.g., within the assumed parameter values). This inevitably brings uncertainty into the answer (or results). The degree of uncertainty is thereby dependent on the least accurate part of the method. In other words, it is nearly never helpful to test one aspect with the highest degree of validity while others with the same sensitivity for the analysis have a lower degree. Measures of quality (such as error intervals) should be given with the final results, or at least the validity of them should be assessed together with the results.

In order to categorize different methods with respect to the subject of evaluation, the model of driver, vehicle, and environment (see Sect. 1.1) can be used. The vehicle is further analyzed using the structure of active safety as given in Fig. 1.6. In practical terms this means that the smallest entity for evaluation within the vehicle is a component. The next level is a system (or some part of a system, here referred to as subsystem). Especially when testing different vehicles, as it is the case, for example, with consumer protection agencies, the levels "function" and "characteristic" are of importance.

As many active safety systems do have a human-machine interface, the driver can also be in the focus of evaluation. The surroundings of the vehicle constitute important influences during development and testing. Evaluation methods refer to multiple possible combinations of different parts of this model (e.g., testing the driver *and* the vehicle or a single component).

The driver with his vehicle and its immediate surroundings form one entity in traffic, but this entity is not isolated. Evaluations often analyze the driver/vehicle entity (or parts of it) as if it were isolated. If the interaction with other participants in traffic is essential to answer the research question given, more than one of those entities must be taken into account.

Once the effects of a measure on (parts of) the system "traffic" (i.e., involving more than one entity), are under investigation, two main approaches can be distinguished:

- *Accident*-based evaluation. The effect of a measure in one or many accidents is investigated. The effect on the whole accident occurrence in a particular area, e.g., a country, can be assessed.
- *Traffic*-based evaluation. In this case, the effects on traffic are evaluated, either in a specific subset or, for example, one country. Depending on the sample size and method used, this procedure includes the evaluation of accidents, as they form a subset of traffic.

The main difference, as explained above, is that a representative evaluation on the sum of both positive and negative safety effects is only possible using traffic-based testing. This point will be discussed in its special meaning with every method further below in this chapter.

2.2 System Responses Available for Evaluation

The evaluation of measures, which are active during the pre-crash phase, includes all possible system responses. As those systems are subject to a variety of uncertainties (e.g., due to limitations of the sensors, variability in the situation when making predictions, etc.) they will not work ideally [2]. That means they will produce unintended side effects; together with the intended effects they can be visualized using a classification matrix [3] as given in Table 2.1.

There are two categories of intended as well as unintended responses with respect to the *objective* danger of the specific situation. The intended actions are the following:

Table 2.1 Categorization of possible system responses

		System response	
		Yes	No
Objective danger	Yes	True positive	False negative
	No	False positive	True negative

- *True-positive* action (TP): The system acts accordingly to its specification[1] in a dangerous situation.
- *True-negative* action (TN): The system does not act in a non-dangerous situation according to its specifications.

The *intended* actions are explained straightforward regarding the objective of the system. If necessary, it should do what it is specified to do (i.e., true positive) and otherwise should not act (i.e., true negative). The *unintended* actions are grouped into:

- *False-positive* action (FP): The system acts like in a hazardous situation while in a non-dangerous situation.
- *False-negative* action (FN): The system does not act in an objectively dangerous situation.

The unintended system actions have different consequences: A false-negative action means, the situation is dangerous and the system should act, but does not. This results in a loss of safety benefit in that situation regarding the specification of the system. A false-positive action is not related to a dangerous situation but can provoke a new critical situation, either if the driver reacts incorrectly to the system action or if the surrounding traffic is endangered (e.g., by massive automatic interventions or incorrect driver actions). In this context, false warnings can be regarded as less dangerous, as they need an incorrect driver reaction to be effective for the surrounding traffic; whereas automatic interventions regarding the vehicle controls have to be considered as potentially more critical [4].

The quality of a measure with respect to traffic safety can thus be evaluated using this abstract scheme. The sensitivity (also called right-positive rate (RPR)), defined in Eq. 2.1, gives the conditional probability that a positive (i.e., objectively dangerous) situation is treated by the system accordingly [3].

$$p \text{ (positive reactions | positive situations)} = RPR = \frac{TP}{TP + FN} \qquad (2.1)$$

The specificity, defined in Eq. 2.2 (also called right-negative rate (RNR)), describes the conditional probability that a negative situation is treated correctly by the system.

$$p \text{ (negative reactions | negative situations)} = RNR = \frac{TN}{TN + FP} \qquad (2.2)$$

The complementary quantity to specificity is the false-positive rate (FPR):

$$p \text{ (positive reactions | negative situations)} = FPR = 1 - \frac{TN}{TN + FP} = \frac{FP}{TN + FP} \qquad (2.3)$$

---

[1] The specification of the system includes the definition of "dangerous" as well as the activation thresholds. "Objectively dangerous" refers to the criteria set within the specification. No generally accepted or universally applicable definition of "dangerous" exists.

The complementary quantity to sensitivity is the false-negative rate (FNR):

$$p \text{ (negative reactions | positive situations)} = FNR = 1 - \frac{TP}{TP + FN} = \frac{FN}{TP + FN} \quad (2.4)$$

Other important rates give information on all correctly (i.e., definition of accuracy) or incorrectly treated situations:

$$p \text{ (correct reactions)} = \frac{TP + TN}{TP + TN + FP + FN} \quad (2.5)$$

$$p \text{ (incorrect reactions)} = \frac{FP + FN}{TP + TN + FP + FN} \quad (2.6)$$

When evaluating the overall safety impact of a measure, the medical term "number needed to treat" (NNT) [5, 6] can be adapted:

$$NNT = \frac{TP + FP}{TP} \quad (2.7)$$

The NNT describes the number of necessary system actions per correct action. Combined with the consequences of false positives, a trade off regarding the overall safety effects can be made. Obviously, NNT is always > 1, but should be as small as possible.

Two concurring ways of optimization are generally predominant during the development of active safety functions and are known as "warning dilemma". The first aims for the highest safety benefit. That requires early system actions as well as high sensitivity. As described, this leads inevitably to an increased number of false system reactions (resulting in lower acceptance as well as possibly new critical situations), as evident in the false-positive rate [7, 8]. The second aim is a low NNT. Optimization which brings down unintended system reactions usually also affects intended system reaction negatively, e.g., more conservative activation thresholds minimize false activations but also lead to later and/or fewer activations. The optimization must thus aim for an optimal trade off with respect to safety benefit, acceptance of the measure, and negative consequences due to false system actions while defining the operating point of the system.

If the consequences of false-positive warnings as well as of false-positive interventions of a specific deceleration could be quantified (e.g., by subject experiments), a factor comparing warnings and interventions could be constructed. For illustration of the methodology, the coefficient "effective intervention" is defined as sum of interventions and warnings, combined by a factor resembling the severity of the "consequences" of each measure. The NNT using effective interventions gives the overall functional "costs" of a system including a combination of warning and automatic intervention. It can be calculated for every desired outcome category (e.g., effective intervention per avoided accident).

The overall functional costs of a specific system configuration as characterized by NNT is one important parameter for optimization. Usually, these costs are intended to remain within a given range. The change in overall costs depending on the optimization parameter is often non-linear. For example, an increase in the time-to-collision, as one defining parameter for intervention by the system, usually leads to an accelerated increase in overall costs, as more and more false-positive interventions will occur per true-positive intervention.

The well-known concept of marginal benefit describes the maximum amount someone is willing to invest for an additional unit of benefit. The marginal functional costs can be interpreted as the derivation or slope of the overall NNT curve. In the case of preventive pedestrian protection, marginal costs refers to the additional cost for another increment of the optimization parameter. In combination with the overall functional costs, a stakeholder or group of them (e.g., manufacturer, driver, society) could set a limit for the overall costs as well as for the marginal costs. The overall functional costs thus narrow down the potential range of the optimization parameter. The optimum within this range could be a minimum NNT (as described above) or an incremental search for the best operating point using marginal functional costs. The parameter is incrementally increased within the range until the marginal costs (i.e., the costs for each additional increment) reach their limit.

An optimized development therefore takes these metrics as criteria for optimization and considers both expected safety benefit as well as possible negative consequences. In order to test false-positive rates or calculate NNT, adequate testing methods with respect to real traffic and its variability are needed [9, 10].

2.3 Retrospective and Prospective Evaluation

Methods for evaluating vehicle-based safety measures can be categorized into *prospective* and *retrospective* [11–15]. The main difference is the time of the evaluation regarding the development process and/or life cycle of the measure in question [12]. Prospective analysis can be used from very early stages of development on (without the necessity of having a fully developed measure), and retrospective analysis can be used once a measure has been developed (and usually has already been in the market for a given span of time) [15].

Retrospective analysis mainly uses real accident data and evaluates existing measures with respect to a safety statistic. A common procedure is to define two groups in the accident data, one with the measure in question and the other without. The two groups are then compared searching for changes in characteristic values of the statistic [14, 15].

There seems to be a consensus in the literature concerning the "power" of this approach as being both very important [14] and impressive [9]. Analysis of existing real-world accident data is even sometimes regarded as "[a]n ideal method to assess the safety impact of advanced safety technologies" [10]. The most prominent example for retrospective evaluation is the Electronic Stability Program (ESP)

[9, 16]. A summary on available studies is included in [16]. Besides the obviously striking approach of using accident data retrospectively for evaluation, past studies also indicate the challenges coming with this method. It took years before the effectiveness numbers regarding ESP turned out to be stable, while other (common) systems like Antilock Braking System (ABS) are still being discussed regarding their actual effectiveness [16].

The retrospective approach in general has a number of constraints:

- The measure must be frequent enough in the market to have a sufficient market penetration and thus produce visible effects in accident data [9, 14, 16–18]. This often takes years as market penetration is dependent on the take rate of a measure (if optional) [10, 11, 15, 16]. The positive exception was again ESP; rapid and broad market introduction lifted this measure quickly above the statistical noise in accident data [9].
- The presence of the measure in a vehicle must be identifiable in accident data [16] in order to group the accidents. The information as to whether an active safety system was active during an accident is rarely available in nearly all accident data sets (this applies only to measures which can be deactivated by the driver). (The limitations of accident data bases in general are discussed further below.).
- Statistical similarity of case and control group must be assured [16].
- Long term behavioral effects may change results over time [16] (see also previous points).
- The retrospective statistical analysis of accident data tests mainly for correlations. The observed effects need therefore not be causally related. A causal relationship has still to be proven, e.g., by controlled experiments [16].
- The baseline or reference group may be biased by avoided accidents, as they are not included in accident statistics [12–14, 16, 17]. Although the opinion exists that only accident mitigation can be evaluated by a retrospective approach [12], newer research indicates that the avoidance potential of measures of active safety may be accounted for by statistical means like odds ratios and thus making this constraint less severe [19].
- Probable interaction effects with other measures can mask the investigated effect: To this end, possible confounders have to be known and controlled (e.g., belt-use rates, presence of other systems with similar functions and/or effects) [16, 19]. The control of confounders is only possible, if those are available in the data sets used. Interaction effects and confounders are even harder to control for, if the data cover a large span in time and the internal influences on traffic and/or accidents may change in that period [12].

The importance of controlling confounders and interaction effects shall be pointed out by an example. Using the retrospective approach, one study evaluated the effects of xenon headlamps on accidents in Germany [20] on basis of the federal accident statistics. As a result, introducing xenon in 100 % of passenger vehicles would lead to a decrease in 6 % of all accidents and 18 % of all fatalities. The study claims that all possible confounders were taken into account and do not bias the results [20]. The possible confounders cited, such as exposure time of vehicles, driver behavior etc., are not part of the federal data, thus cannot have been accounted for in the study. The results found may not be attributed to xenon headlamps as stated but could be causally connected, for example, to differences in driver behavior (as xenon was introduced in upper class vehicles first, they could have a different driver population). The vehicle groups could be very inhomogeneous if differentiated by xenon, as the comparability of drivers and vehicles cannot be assured in the groups used. All other safety features of vehicles introduced during the 10 years of data considered (coded in the federal statistics or not!) were not explicitly taken into account. This example should highlight the importance as well as the difficulties considering the challenges while performing retrospective analysis as described above.

The necessity to evaluate safety measures *before* market introduction regarding their safety benefits is the driving force behind *prospective* analysis and evaluation [12, 21]. In this case, only one group is selected from the data instead of two. This group is then evaluated on a theoretical basis with and without the measure in question [14].

The advantages of prospective analysis in general are the following:

- Applicable from early on during development of a measure [9, 14].
- Using one group only eliminates several problems stated above for retrospective analysis (e.g., comparability of the two groups) [14].
- Possibility to compare different variations of a measure during development [21].

The limitations of prospective analysis are not that easy to generalize, dependent to a large extent on the method used. These methods have a wider variation than the retrospective ones. Different examples are given in the following sections together with their specific advantages and challenges. The main challenge for any method is its validity with respect to the question it tries to answer.

2.4 Data Sources for Evaluation

As both retrospective and prospective evaluation methods are based in many cases on accident data, a short summary concerning possibilities and limitations of accident data as well as other data sources is given in the following. These general findings have effects on the validity of each method discussed below and are not dependent on the specific method used.

There is a variety of different accident data bases available for evaluation. Two main criteria for categorizing accidents data bases are representativity and level of detail [12, 21]. The representativity is directly but not entirely linked to the number of cases available in the data set. Another factor is the representativity of the sampling scheme used. As a consequence, two categories of accident data are in-depth and national (or international) data collections.

National statistics are regarded as being most representative for their specific country. For example, the German Federal Statistics, provided by the Federal Statistical Office (Statistisches Bundesamt), collects all police reported traffic accidents connected to driving traffic. That means that police reported accidents involving only pedestrians are excluded from this statistics. As the police are mainly contacted in case of personal injury or high property damage, accidents with slight injuries or minor property damage may be underreported [21, 22]. The federal statistics have high case numbers, but also a low depth in the data, as all information is taken from police reports. Access to disaggregated data is limited [16, 21]. Especially information regarding the genesis of an accident, the course of events during an accident, the vehicle damage in detail, and the injury mechanisms are not included [21].

Also on national level, the German Insurance Association (Gesamtverband der Deutschen Versicherungswirtschaft e.V., GDV) runs its own accident data base. It contains detailed documentation of a sample of all accidents followed by insurance claims. The main sampling criteria are personal injury and a property damage greater than €15,000 [23, 24]. Around 700 cases are added each year [24]. Access to the database is limited to the members of GDV [13]. Although the representativity for accidents with insurance claims within these criteria is given [23], a further extrapolation of the findings is difficult due to the biases induced by the sampling scheme [13].

In-depth accident investigations include more details but contain by far fewer cases [16, 21]. One example is the German In-Depth Accident Study (GIDAS). As a combined industry and government effort, about 2,000 cases are collected and documented with a high level of detail each year [25] (see also Sect. 5.2.1 for a general description of GIDAS). The main sampling criteria are accidents with at least one injured person in road traffic. The sampling area is confined to two German cities and the surrounding areas; the sampling itself follows a shift schedule [25].

As a consequence, some restrictions apply when discussing the validity of findings based on in-depth accident data. The restrictions given in the following refer to GIDAS as an example, but can be transferred to other studies with respect to their internal structure and sampling criteria:

- Only accidents with personal damage [14, 21]; thus severe accidents are over-reported.
- No accidents with property damage only or non-collisions (i.e., critical situations) [21].
- No information about participants in traffic who were not directly involved in the accident [21].

Other biases may be induced by low case numbers as well as other sampling criteria [16].

The results based on accident data are only valid for the area of the data set [16]. Nevertheless, generalization of the findings, for example, to national level, is facilitated using weighting procedures. Based on parameters available in the national statistics, a weighting scheme tries to correct biases in the GIDAS data and thus make them more representative for Germany. The most commonly used scheme relies on type of the accident, accident severity, and location of the accident (urban or non-urban). A description of the procedure can be found in [14]. In order to gain representativity, this weighting or very similar approaches are widely used [7, 13, 16, 25].

Officially, weighting ensures that GIDAS is mainly representative for the areas its data is collected in as well as for most aspects of passive safety, if free from regional influences [25]. However, the benefits of weighting as well as the validity of the results are still subject to discussion. On the one hand, this procedure is believed to ensure representativity [26]. On the other, studies show that weighting does not solve these problems and is not able to correct all biases in the data set. Even sophisticated methods still leave distortions in the data [27, 28]. As the representativity for the country of the in-depth study itself has to be questioned, an extrapolation on other countries seems even less valid; in this case, the use of accident data directly collected in that country is recommended [21].

Another example for an in-depth accident data base is the Pedestrian Crash Data Study (PCDS) from the US [29] (which is also described in Sect. 5.2.1) or accident investigations carried out by vehicle manufacturers. The latter ones have a very high level of detail but suffer even more from biases due to low case numbers, model selection criteria or geographic effects [16].

For more information and the description of different accident data bases, also on an international level, the following literature provides a good starting point in form of summaries [13, 14, 30, 31].

Comprehensive and detailed knowledge of all factors relevant in accident genesis are a prerequisite for an evaluation of safety during the pre-crash phase [32]. A variety of factors is available in accident data bases (see above), but many factors—especially relevant during the genesis of an accident—are not part of accident data bases [32, 33]. As a consequence, detailed conclusions about the pre-crash phase and the genesis of an accident, especially with respect to critical combinations of mistakes and the course of events following the phases of an accident, are only possible in a very limited way [32]. Thus, the understanding of the mechanisms and processes involved is also bound to these limitations [33]. As many parameters, especially concerning the persons involved (e.g., the driver), are not available in accident statistics (and cannot be gathered by methods applied in accident data collection) [34], a distinction of different accident causes is very difficult [35].

The reliability and validity of the accident data proves to be a difficult issue as the data collection is always a sample and not a census in the sense of an absolute "true" number [33, 35] (see also abstract description of different accident bases and sampling schemes above). Furthermore, even the data available most of the time include inconsistencies and uncertainties due often to the process of reconstruction and the assumptions necessary therein [35].

Although accident statistics are able to give valuable information about accidents as well as influencing factors (at least to some extent), the findings must still be interpreted with care, as their true meaning is only revealed when related to exposure [32]. Many studies using accident data do not consider risk exposure or discuss the correct measure for exposure with respect to the research question [34].

Accidents are statistically rare events [17, 32] and represent only partly the complexities of traffic. As discussed in Sect. 2.1, they cannot be regarded as being sufficient for every possible evaluation of safety in traffic [33], especially of the pre-crash phase. The events leading up to a possible accident are by far more frequent than the accident itself [36] (see also Sect. 1.2). From the whole course of events from "normal" driving, mistakes, failed corrections, contributing factors, and finally a collision [37], only the last part is recorded in accident statistics. Consequently, no data on avoided accidents or very slight accidents are in the data bases [17]. The evaluation of safety benefits using accident data as the only data source is thus regarded as incomplete [17]. The evaluation of overall safety effects with respect to false activations etc., as explained in Sect. 2.2, is also not possible solely on the basis of accident data.

Directly linked with the "systematically missing" information in accident data bases is the importance of the human behavior for the accident genesis and thus for the evaluation of measures of the pre-crash phase. In most cases, human behavior is by far more important than driving performance of the vehicle [17]. The driver has a highly variable behavior and is the decisive element during the pre-crash phase (as well as during "normal" driving) [38]. Measures that interact with the driver—by direct means of a human machine interface (HMI) or indirectly via their interventions in vehicle controls—require a driver model for their theoretical (i.e., not subject-based) evaluation [13]. In case an HMI is present, it has to be evaluated, too [23]. Due to the importance of the driver [39] an over-simplification of the driver model leads to invalid results. The complexities of human cognitive modeling are avoided in some studies using a "perfect" driver (often with fixed (re)actions, not distributed stochastic behavior) which constitutes a severe assumption [23, 24].

In this context, not only the driver in view of his actions and decisions is of interest, but also regarding his acceptance of different measures [39] (see Sect. 4.3 for more on acceptance and its connection to safety). During the use of a system of active or integral safety, changes in behavior due to adaption or compensation effects can occur and must be accounted for when evaluating safety benefits [32]. To draw a conclusion, a "full forecast of their [i.e., active safety systems; author's note] potential is only possible with respect to the complete relation of driver-vehicle-system-environment" [9] (see also [40]). These and various other aspects of evaluation of active safety systems have also been subject of discussion in European Union funded projects; an overview is, for example, given in [41].

The limitations and boundary conditions connected with *accident-based* testing lead directly to the approach of *traffic-based* testing, based on the control loop of driver, vehicle, and environment with its processes on the way to an accident [42] (see also Sect. 2.1). The main advantage of traffic-based testing over accident-based testing is that exposure can be taken into account [43]. The usage behavior of the driver regarding active or integral safety systems can also only be reliably tested in real traffic [44]. Different methods considering the variability of traffic and its consequent testing are discussed below. Traffic-based testing does not necessarily mean driving on public roads but also can involve simulation [10]. In addition, traffic-based testing is capable of evaluating all four possible categories of system responses (see Table 2.1), where accident-based testing is mainly restricted to evaluating true positives and false negatives. As false positives as well as true negatives are of high importance while finding the optimal operating point for a system, traffic-based testing is the method of choice and is able to close that gap.

2.5 Methods of Prospective Evaluation

This section discusses a variety of rather "theoretical" methods of prospective evaluation: fault tree, traffic conflict technique, operational field and effectiveness in the field, and scenario technique. Section 2.6 focuses on case-by-case prospective methods.

The basic idea of analyzing the traffic system with respect to the genesis of mistakes, conflicts, and accidents is structured in the *fault tree* method [45], see Fig. 1.1 (p. 3). This scheme follows a process-oriented approach as displayed in Fig. 1.2 (p. 4). The basic concept of Reichart was thus extended beyond the basic elements of driver, vehicle, and environment to include, for example, driver assistance systems [42, 46]. One advantage of a fault tree is that once a critical set of probabilities is known, the calculation of the other probabilities is straight forward using the Boolean connections in the tree.

In this context, the top event in a fault tree can be an accident or a conflict [45]. The probabilities for accidents can come from classic accident analysis [42], the corresponding ones for conflicts or mistakes (being at the other end of the tree structure) are not generally known and hard to extract [35, 42]. An example of such calculations as well as further information, for example, on validity of the method, can be found in [45]. The method seems to be able to generate sound results, especially on the connections between conflicts and accidents, although many assumptions are basically needed during evaluation [36].

As conflicts can be top events in fault tree analysis or generally constitute rather high level events, the probabilities and nature of conflicts is regarded an important issue within the literature. One way to assess conflicts in traffic is the so-called *traffic conflict technique* [35]. A traffic conflict may be characterized by considering approaching object trajectories which, extrapolated in time, would exhibit an increased probability for collision unless one of the participants changes his current state of motion [35]. This definition could be extended on non-observable situations and single vehicle conflicts.

The objectives of this standardized observational technique are risk assessment as well as effectiveness evaluation of traffic facilities, not estimations regarding the quantity of accidents [35]. Thereby, conflicts have a probability to become accidents, which does not mean that accidents can be predicted with the method [35]. The transition probabilities between conflicts and accidents, as needed, for example, in the above-mentioned fault tree analyses, can be assessed [42]. Compared to accident analysis, investigating conflicts has the following advantages [35]:

- Conflicts occur with higher frequency and thus provide more statistical power.
- Conflict data can be collected with more completeness and better controlled reliability.
- They allow for a more "objective" collection, as legal liability is not considered.
- Conflicts have still sufficient frequency even for low accident frequency at that point.

- Regional boundaries as well as other requirements of data collection can be well defined and documented.
- Conflicts can be collected as a controlled sample.
- Many additional factors can be collected.

The studies done show that conflicts have a good correlation to accidents and thus can be considered as "dangerous" [35]. Therefore, the results (respectively the probabilities) can be included, for example, in fault tree analyses. However, the traffic conflict technique itself requires a high effort in data collection and analysis [45]. Further combined research on mistakes, conflicts, and accidents is strongly recommended in the literature [45].

Which specific factors can possibly be investigated using traffic conflict technique, depends on the technology used (in general, a large variety of factors are well observable in traffic). A *static* traffic observation from a fixed location may adequately record all macroscopic traffic effects as well as environmental parameters, but has its limitation regarding the precise measurement of dynamical properties of individual participants as well as their specific configurations (i.e., presence of specific safety measures, such as ESP). The technique of traffic observation from the view of one specific participant will be discussed in connection with Naturalistic Driving Studies (NDS) and Field Operational Tests (FOT) further below (see Sect. 2.7).

Another possible evaluation uses the *operational field* and the *effectiveness in this field* of a measure as a metric. Two commonly used definitions exist:

- The operational field (OF_1) is defined as the number of accidents, where the measure can potentially show an effect:

$$OF_1 = \frac{\text{potentially affected accidents}}{\text{all accidents}} \tag{2.8}$$

The effectiveness (EF_1) within this operational field is then defined as the quotient of real effectiveness to the number of accidents [13]:

$$EF_1 = \frac{\text{affected accidents}}{\text{all accidents}} \tag{2.9}$$

- Another definition is given by the following equations [14], where $OF_2 = OF_1$.

$$OF_2 = \frac{\text{potentially affected accidents}}{\text{all accidents}} \tag{2.10}$$

The effectiveness is given as:

$$EF_2 = \frac{\text{avoided accidents}}{\text{potentially affected accidents}} \tag{2.11}$$

Considering all accidents (e.g., in one country or area) as baseline, the *overall effectiveness* [14] is given by the product of 2.10 and 2.11:

$$\text{effectiveness}_{overall} = OF_2 \cdot EF_2 = \frac{\text{avoided accidents}}{\text{all accidents}} \qquad (2.12)$$

The advantage of the first definition of effectiveness (Eq. 2.9) is that it includes all possible effects of a measure, also negative ones. The second definition (Eq. 2.12) refers to positive (intended) effects only. The effectiveness in general is assumed to be smaller than the operational field, as no system works perfectly in the sense of an ideal system [14].

The advantages of this method are that it allows a fast application with limited effort. The operational field can be estimated quite exactly (e.g., using accident data) whereas the effectiveness estimation is a challenge and is often facilitated by assumptions, resulting in low validity [13]. Although this method has become a common practice in the last years, the procedure itself was used decades before.

One example is a summary report on the research that led to the introduction of the center high mounted stoplamp from 1985 [47]. The overall effectiveness in the sense of Eq. 2.12 for the center high mounted stoplamp was determined by operational field and effectiveness. The operational field was defined as the number of all rear end accidents. The effectiveness was estimated from different studies (most of them FOTs) and has to be regarded as much more valid than, for example, an expert opinion. The average effectiveness was found to be 50 %. In addition, an overall monetary benefit was calculated, first on the basis of avoided accidents, secondly using a cost-based (monetary) approach for both avoided and mitigated accidents. As a result it was possible to give a cost benefit ratio for the measure, which was found to be 0.1 [47].

In line with the previous approach is the *scenario technique*. It describes the possible benefit of a measure regarding accidents of relevance [13]. An exact effectiveness is not determined, but the true effect is approximated using two scenarios as upper and respectively lower boundary. The scenarios are defined using an optimistic and pessimistic approach with respect to the benefit [13, 14]. The analysis is commonly conducted using accident data and assumptions on the effect of a measure.

The methods mentioned above can rely on assumptions to a particular extent. One very common form of making assumptions are expert opinions. Although the general value of expert opinions should not be a matter of doubt in this work, the validity regarding the effectiveness of complex systems in complex (and highly variable) traffic or accident situations has to be doubted. Depending on the extent and severity of the assumptions used, the validity of a study has to be questioned. For example, if the whole effectiveness of a measure is based on expert opinion alone and is not backed by any empirical evidence, then this constitutes a severe assumption. In order to demonstrate a method [14], this can be regarded as uncritical, but in real evaluations this should be avoided, at least regarding sensitive parts or models of the evaluation process.

2.6 Methods of Prospective Case-by-Case Analysis

The accident-based methods described in the previous sections use aggregated data. That means that coded values of many accidents are used to define, for example, an operational field for a measure without considering each single case in detail. These procedures have the disadvantage that subtle effects (e.g., interactions between a system and a driver or the perceptibility of an object by a given sensor over time) can be considered only in very general means on a meta level. The potential of a future measure can thus be evaluated at a very early stage of development. However, optimization and evaluation of minor system changes are not possible on that level but require more detailed analysis with respect to the time-bound interactions in each single case.

Although single-case analysis is not new and has been conducted for decades in different ways, modern calculation capabilities together with corresponding detailed data sources allow a very detailed analysis which is not limited to a few cases any more [13]. As a result, a large number of cases can be analyzed automatically with reasonable resources [13, 14]. This kind of analysis solves the problems of time consuming and complicated hardware testing in many different situations and thus is reproducible without danger, quantifiable, and controllable [48]. Ideally, a flexible and universal tool would fulfill those characteristics instead of an inhomogeneous world of incompatible tools [11]. All relevant parameters should be adjustable and the boundary conditions variable in order to enable sound testing and evaluation of the safety effects [11].

Considering simulation as a method, the validity of the findings is a key aspect. The simulation itself must be validated regarding the research question it is used for [49]. In addition, a validation and, of course verification, if possible, of the findings against field data (e.g., accident data) can also be recommended [11].

The following part of this section briefly introduces different methods that can be categorized as prospective case-by-case analysis. Case-by-case evaluation is explained using the injury shift method as example. Different methods including case-by-case simulation are described:

- Simulation by Busch
- PreEffect-iFGS
- rateEffect
- VUFO Simulation
- PreScan
- Bosch simulation
- ACAT simulation

The first one focuses on the evaluation of passive safety measures and is called the *injury shift method*. The basic idea is that a passive safety measure has a positive effect on the severity of injuries sustained at a specific component. The assumption is that below 40 kph, optimized components result in a reduction of one level on the abbreviated injury scale (AIS) [50] (for a detailed description of AIS see Sect. 5.1,

p. 91). As a consequence, this may lead to a reduced overall injury severity. The benefit of a given measure is thus evaluated on the level of single injuries and corresponding components in each single case [14, 51]. The method has been used in a couple of studies [14, 30, 51–54].

The injury shift method has been used, for example, for the evaluation of secondary safety measures for pedestrians at passenger cars. The maximum impact speed considered is 40 kph. The results of Euro NCAP crash tests are transferred to the vehicle in question, and the impacting body parts are mapped to the test grid. The metric includes assumptions leading to an optimistic and a pessimistic approach. In the optimistic case, if the zone was tested green, the injury severity is shifted down to AIS 1 (pessimistic: by two AIS levels). If the zone was tested yellow, the injury is shifted by two AIS levels (pessimistic: one AIS level). A red zone does not lead to a shift in either approach. No injury is shifted below AIS 1, meaning the method does not predict avoided injuries [53].

The injury shift method is computationally efficient: The assumptions used lead to an algorithm, which is simple and fast to calculate and can be applied to a table of injuries and corresponding vehicle components. Each case is thus evaluated, and the overall injury severity of every person is recalculated. As a result, the safety benefit for every person can be evaluated in comparison to the original severity distribution.

However, several severe assumptions underlie this estimation method: As detailed Euro NCAP test results are not available for the majority of (rather old) vehicles in databases such as GIDAS, each vehicle is considered as zero points (i.e., being totally red). In reality, also older vehicles do have a good protection potential in some zones and the overall safety benefit of a measure is thus overestimated by the underlying assumption [21]. The second challenge is that a color in the Euro NCAP test stands for a bandwidth of actual dummy readings. That means a color distribution is a rough estimate of the real stiffness (protection potential) of a vehicle. The three-color categories used for the injury shift method can be regarded as rather crude approximation to a stiffness distribution. In addition, all AIS levels are treated in the same way without considering that AIS is a non-metric scale. It is unclear whether a given measure has the same effect on an AIS 2 as on an AIS 5 injury [21, 52]. Considering pedestrians, only the impact on the vehicle is evaluated, not the secondary impact.

The next level in automated single-case analysis is the actual simulation of the dynamics over time for each accident. The focus is on the pre-crash phase of an accident. In 2005, Busch described a simulation of single accidents, each with and without the measure of active safety in question [13]. The main procedures are: selection of relevant accidents, simulation with/without system, translation into injury severity, and calculation of the effectiveness. The input data for the simulation are the values coded in the GIDAS data base. As the sequence of the accident is described there via characteristic parameters but not as time series, the simulation provides a kind of automated reconstruction of the pre-crash phase and a subsequent simulation of it. By comparing the results for each accident and summarizing them, the effectiveness is calculated. The first stage is a physical assessment (i.e., impact speeds, impact locations, etc.). These data can be translated into physiological data using, for example, the injury shift method for passive safety and injury probability models for active safety (for the later see also Chap. 5). The results gained from the simulated accidents are then weighted to the national statistics to gain representativity for Germany [13].

The advantages of this approach are a degree of representativity as well as the opportunity to model a system in detail and take system modifications into account by simulating each single case. A drawback of the method is its reliance on the information available in the data base used. As many relevant pre-crash parameters are not coded (and especially not coded as time series) in in-depth data bases (such as lane markings, positions of the vehicles in the lanes) and thus are not available for the simulations, only a limited subset of functions can be evaluated (e.g., automatic braking, but not lane departure warning). The method does not include behavioral driver modeling, i.e., models of driver perception, response, performance under extreme condition, etc. These are required for an evaluation of a system with an information or warning component. Thus, this method is limited to evaluation of automatically intervening systems [21].

The next evolution of the method presented by Busch is called PreEffect-iFGS. It is a prospective method for evaluating the field effectiveness of integral pedestrian protection systems [21]. The main procedures of Busch, i.e., selection of relevant accidents, simulation with/without system, translation into injury severity, and calculation of the effectiveness, stayed the same with some additions. The improvement is an incorporation of test results for active and passive safety systems derived from hardware testing [54]. The initial version also includes an automated backwards simulation of each accident based on the values available in GIDAS. The results are then transferred into the commercial software PC-Crash and are then simulated forward with and without the measure in question.

The simulation can be run in two modes: open-loop and closed-loop. The *open-loop* variant calculates key parameters for automatic interventions with different parameters per accident. These key parameters are then filtered using the specific system configuration in question. The advantage is that a variety of system configurations can be compared without running the simulation again. The disadvantage is that the results do not include a feedback loop of the measure on the situation itself, for example, the reaction of a driver to a warning. The *closed-loop* simulation includes the feedback on the situation and thus is able to evaluate all kinds of effects, e.g., the driver's reaction to a warning. The higher level of detail and the inclusion of a probabilistic driver model increase the computational effort [21].

One main disadvantage for the simulation methods described above is the inherent limitation regarding depth of information of the data used. In order to make more information during the pre-crash phase available and thus enable other functions to be evaluated, a project has been launched within the GIDAS consortium. The so-called *pre-crash matrix* is a digital and machine readable description of the pre-crash phase [55, 56]. The information falls into the categories *static* and *dynamic*. The *static* part contains information on the street layout, the lane markings, and accident relevant objects (e.g., parked vehicles). The *dynamic* part contains the trajectories of the participants as a time series, going back about 3 s before the first collision.

The information ends at the point of the first collision [55]. This data base provides a uniform basis for simulation of a subset of the GIDAS accidents, thus making the backwards simulation of accidents as used in the method above obsolete. In addition to the pre-crash matrix, values from the GIDAS data base, such as vehicle characteristics, weather conditions, etc., can be used.

The next version of PreEffect-iFGS, called rateEffect, is able to import the pre-crash matrix and use these data [57–61]. Whereas rateEffect as well as a comparable approach from Spain [62] use PC-Crash as software package, other solutions are available, too. PC-Crash is able to calculate the crash phase and thus rateEffect provides key parameters of the crash (if still one occurs) as additional input for an evaluation metric.

The Verkehrsunfallforschung an der TU Dresden GmbH, which is one of the data collecting partners in GIDAS, developed a pre-crash simulation using a commercial driving dynamics simulation as core together with proprietary Matlab® and Simulink® functions [55]. The latest version changed the driving dynamics simulation from CarSim™ to CarMaker™ [63]. The idea is again to simulate single accidents automatically and to compare a system effect to a baseline without system. Also in this simulation, no predefined field of operation or estimated effectiveness is needed [55].

The Netherlands Organization for Applied Scientific Research TNO has developed another simulative approach called PreScan®. It includes the complete road situation, vehicle sensors, system controls, and vehicle dynamics [64]. Based on Matlab®, Simulink®, and Stateflow®, PreScan® claims not only to simulate the pre-crash phase, but also to calculate the crash consequences via a link to MADYMO® [48].

The Robert Bosch GmbH developed a Matlab®-based simulation working with GIDAS accidents [39]. One essential part (especially for systems with a human machine interaction, such as warnings) is the modeling of the driver in terms of cognitive processes. The cognitive modeling of the driver is also capable of revealing findings about system acceptance and thus effectiveness (see also Chap. 4).

All methods described above can be categorized as automated case-by-case simulations based on accidents. There are two more aspects which are of importance for a sound system evaluation during the pre-crash phase. Many processes involved are deterministic, e.g., the participants dynamics, the technical functions implemented, as well as many physical boundary conditions. However, some of the key processes do have a stochastic nature; for example, the driver action and reaction as well as some characteristics, e.g., of the sensors modeled. Due to the sensitivity of the results to those processes, stochastic elements are an important feature of any representative evaluation (see also Sect. 3.4).

For example, the driver reaction is important for the genesis of an accident as well as the interaction with a safety system and the possible impact of a safety system. As a consequence, stochastic driver modeling is also included in some approaches [15, 65]. Stochastic elements are not limited to processes within an accident but are of importance also in uncritical traffic situations. As mentioned before (see Sect. 2.2), an overall estimation of possible safety effects should include the evaluation of positive effects within accident scenarios as well as undesired potentially hazardous side effects in normal traffic. The only data source used in the approaches discussed above, i.e., accident data, does not provide normal or critical situations which would not have resulted in an accident. There are several ways to incorporate this traffic-based evaluation into a simulation.

As classical data collections are limited to accidents, one way to get data on non-collision events is a stochastic variation of accident reconstruction data in a way that the single event does not necessarily result in an accident anymore. These non-collisions are then used in the simulation in order to assess the balance between desired and undesired effects of a measure in traffic [15]. As a consequence, validating the non-collisions regarding the distribution of key parameters and their representativity for overall traffic is vital. The basic data concerning exposure are not as well known as accident related data.

Another simulative method is introduced in the following in order to highlight the use of other data than accident data. Within the Advanced Crash Avoidance Technologies Program[2] (ACAT), initiated by the National Highway Traffic Safety Administration (NHTSA), a standardized Safety Impact Methodology (SIM) has been developed [10, 66, 67]. The main objective was to develop a tool that evaluates the effectiveness of crash avoidance technologies in a US context. Combined with that is the development of objective tests capable of verifying the safety impact of a real system [10].

The basic idea again is to conduct time domain-based simulations of the driver-vehicle environment with and without an ACAT system. The available data include crash cases from accident data bases, such as the National Automotive Sampling System (NASS) Crashworthiness Data System (CDS) or the Pedestrian Crash Data Study (PCDS) (for a detailed description see Sect. 5.2.1, p. 93). The difference to the approaches discussed above is that normal driving situations from Naturalistic Driving Studies (NDS), for example, provided by Virginia Tech Transportation Institute (VTTI) as well as synthesized crashes are also used for evaluation; the technique thus allows for an assessment of non-critical situations [10]. Again, the driver reaction can be varied, e.g., using Monte-Carlo techniques, and the simulated physical outcomes are translated into physiological parameters. The results can be transferred to national US level [10, 68]. The ACAT framework is thus an example for an accident as well as traffic-based approach.

2.7 Methods for Modeling Different Parts of Driver, Vehicle, and Environment

This section introduces typical methods for investigating the control loop of driver, vehicle, and environment or different parts of the latter (see also Sect. 1.1, p. 1). Although the methods are not able to assess overall effectiveness of a measure of

---

[2] Extensive information in form of publications and project reports is available on the NHTSA homepage.

active safety, they provide valuable findings that can be incorporated into modeling as needed for the simulative approaches discussed above. This section is not meant as a complete compilation but aims at giving an overview about the different levels of testing in the traffic system as well as typical examples of current methods. Included in this overview are:

- Hardware and component-based testing
- Vehicle Related Pedestrian Safety Index (VERPS)
- Vehicle Hardware In The Loop (VEHIL)
- Test track
- Test track target: Experimental Vehicle for Unexpected Target Approach (EVITA)
- Subject experiment: Driving simulator
- Vehicle in the Loop (ViL)
- Real traffic
- Naturalistic Driving Study (NDS)
- Field Operational Test (FOT)

The smallest unit under testing in the context of active or integral safety is a *component*. This method is usually applied for measures of passive safety, such as deformation spaces, active bonnets or airbags. This "classical" testing can be based on hardware (like the Euro NCAP tests for pedestrian safety [69, 70]) or virtual testing using, for example, multi-body or finite element simulation. These methods are not discussed here in detail (examples can found in [48, 71, 72]) but are relevant for active safety, since the concept of integral safety (see Sect. 1.1) includes a comparison of the effectiveness of active and passive safety. Some of the methods discussed below also incorporate component-based testing.

One example of a component-based testing method is the *Vehicle Related Pedestrian Safety Index* (VERPS) [73, 74]. This index utilizes a linear scale for both active and passive safety measures. The pedestrian head impact in frontal passenger vehicle collisions is assessed using the Head Injury Criterion (HIC) as metric. The method delivers specific results for a given vehicle and pedestrian combination. The evaluation process includes accident data analysis for relevant scenarios, kinematic analysis (via multi-body simulation), hardware component testing, and a procedure to obtain the VERPS index [73, 74]. The VERPS index takes only the probability for AIS3+ head injuries due to impact on the vehicle into account, since this probability can be derived from the HIC measurement.

As an addition to the VERPS method, the (here slightly generalized) *VERPS+$_k$* index considers the effect of active safety, as different impact speeds lead to different kinematics and impact locations as well as changed HIC values.

$$\text{VERPS+}_k = P_{impact}(v) \cdot \sum_{i=1}^{m} \sum_{j=1}^{n} R_{i,WAD}(v) \cdot R_{j,front} \cdot \left(1 - e^{-\left(\frac{HIC_{ij}(v)+500}{1990} \right)^{4.5}} \right)$$

$$(2.13)$$

- $P_{impact}(v)$ gives the dependency of the impact probability for the pedestrian's head on impact speed v.
- $R_{i,WAD}(v)$ and $R_{j,front}$ are relevance factors with respect to the impact probabilities derived via analysis of accident data: $R_{i,WAD}(v)$ refers to the relevance in longitudinal direction, depending on impact speed, and wrap around distance (WAD). $R_{j,front}$ gives the corresponding relevance in lateral direction.
- $HIC_{ij}(v)$ characterizes the pedestrian's head loading, depending on impact speed and the area on the vehicle front, as specified by i and j.

The secondary impact is not assessed, but is assumed to improve with decreasing impact speeds [73, 74]. The actual performance of an active safety system together with the driver (if a driver-relevant component is included) is estimated by weighting the different $VERPS+_k$ indices for different speeds according to the performance of the active safety system (including avoided accidents and the probability of avoided head impact on the vehicle). Averaging over all drivers in the population in question and all relevant accident situations, the resulting VERPS+ index is able to quantify the effect of an active safety system [73, 74] once the primary effect of the active safety system (i.e., the reduction of vehicle speed) has been assessed properly. An example for the use of the VERPS index as well as an addition for leg injuries can be found in [75].

An advantages of the VERPS method is that both active and passive safety can be assessed on a common linear scale. A drawback is that only pedestrian head injuries in primary frontal passenger vehicle collisions are evaluated, while secondary impacts are not taken included [73, 74].

Clearly focused on active safety functions is the *Vehicle Hardware In The Loop* (VEHIL) facility of TNO in Helmond, Netherlands [49, 77, 78]. The basic idea is to connect a traffic flow simulation with a chassis dynamometer for testing active safety systems as hardware including the whole vehicle. The surrounding traffic is represented by moving platforms as in Fig. 2.1, which can be fitted with shapes and materials suitable for the specific sensors used. As the vehicle under investigation is on a dynamometer, the moving bases just have to perform the relative movements to the (not moving) vehicle. VEHIL is intended for testing, for example, Adaptive Cruise Control (ACC), collision warning systems or functions based on car-to-car communication [77]. The advantages of VEHIL are the possibility for safe, reproducible testing with real objects. In addition, the actual state of all participants is known and can be analyzed [49]. The limitations are, for example, a minimum time-to-collision about 0.5–0.2 s and a maximum relative speed about 50 kph [49].

A *test track* is a "classic" environment for testing and evaluation of different functions [79]. A experiment on a test track can reproduce very different aspects of various traffic systems, such as different kinds of road classes, road surfaces or traffic situations. As test tracks are not open to normal traffic, full experimental control [44] together with a quite realistic environment [80] including real vehicles and their dynamics is available [79]. Another advantage is that test tracks are available on many locations around the world which makes testing geographically flexible [80].

Fig. 2.1 VEHIL test facility: **a** vehicle under investigation on **b** dynamometer together with **c** moving bases [76]

The challenges come with the construction of specific traffic situations on a test track. Some situations are hard to build (such as complex ones with many participants) or are dangerous (especially for safety related functions) [49]. This leads to two consequences: First that it requires a high effort on a test track to build a subjectively critical situation which is objectively safe; and second that scenarios have to be kept quite simple and perhaps must be within a limited speed range [80]. As test tracks lack normal traffic situations, testing can be less diverse and realistic than in road traffic [44].

Although test tracks provide a valuable environment for development and testing, "a test track test alone will not be sufficient" [9]. Concerning the possibilities on a test track it can be concluded that "it is not realistic that [...] overall functionality and performance [can] be evaluated on basis of a limited number of tests" [9]. For overall effectiveness of a measure of active safety, track testing alone does not seem to be sufficient [10]; thus statistical methods or field tests seem to be more promising [9].

Due to the importance of hardware testing on test tracks during development, a common practice for the evaluation in safety-critical situations is introduced here. In order to achieve a subjectively realistic but objectively uncritical situation for active safety functions, so-called *targets* are used on test tracks instead of real traffic participants. Whereas most targets are designed for sensor or system testing, some are also suitable for behavioral studies. Many targets represent vehicles, or what a sensor or driver can perceive of a vehicle. For radar this means that a triple reflector made of

Fig. 2.2 EVITA: lead vehicle and trailer [86]

the right material can be sufficient, whereas for a mono-camera, a picture of a vehicle is sufficient. Important features are the possibility of self-propulsion, the ability to be crashed (without damage to target and vehicle) and the sensor characteristics of the target. Descriptions of a large variety of different targets are found in the literature [8, 49, 81–84]. Not only vehicles, but also, for example, pedestrian dummies (targets) fitting specific sensor requirements are available [8, 10, 85].

Representative for the variety of targets used in research and development, a more advanced target capable of performing system as well as subject experiments (including behavioral as well as acceptance studies) is introduced in the following. The *Experimental Vehicle for Unexpected Target Approach* (EVITA) has been constructed for testing critical rear-end situations and aims for a high degree of reality [82]. Figure 2.2 shows the trailer with an original vehicle rear-end connected to a lead vehicle via a cable and a winch in the lead vehicle. The trailer is comparably lightweight and can be braked independently of the lead vehicle, thus suddenly reducing the gap to the following vehicle with the system and/or subject on board. The trailer looks realistic for the subject (including full brake lights, etc.) as well as for many common sensors. If the time-to-collision (TTC) reaches a defined value (measured by a backward radar sensor in the trailer), the winch closes, and the trailer is accelerated away from the following vehicle [83]. This allows safe and reproducible testing under quite realistic circumstances [82]. EVITA is limited to rear-end situations and is not impact resistant, allowing a minimum TTC of about 0.8 s and a maximum relative speed of 50 kph. Its velocity is limited to about 80 kph [83].

The importance of *subject experiments* (or behavioral studies) is founded on the fact that for active safety, driver behavior is more important than the driving characteristics of the vehicle [17]. However, human behavior is subject to a large variability [43], which can be modeled, e.g., on the basis of experiments [87]. The findings from many experiments can then be used to develop behavioral models [87] which, for example, are used in simulations as described in the preceding section. The EVITA target for use on a test track is one possibility for assessing driver behavior in particular situations.

A common practice for the evaluation of active safety is use of *driving simulators*. A driving simulator provides an environment for subject experiments with the aim of assessing usability and ergonomics, for example, of advanced driver assistance systems (ADAS), investigating driver behavior with (and without) a system in different situations, and generating findings on acceptance [79]. Driving simulators can have a variety of setups and functions. They range from simple static mock-ups (which basically include a display and human-machine interface) to dynamic simulators, which can simulate limited motion with respect to six axes, allow a full range of view, and provide the look and feel of a real vehicle for the subject [44, 49, 79].

The main advantages of driving simulators are high reproducibility of experiments [18, 79, 80] together with very good experimental control [80]. The possibility of collecting detailed data (including, for example, the surrounding artificial traffic) provides the basis for comprehensive analysis [18].

Experiments in the driving simulator can be conducted during a very early stage of product development [18]. The safe testing of critical situations without endangering the subject allows the investigation of functions which are in early development phases and which thus are not sufficiently safe, e.g., for test track experiments with subjects [18, 79, 80]. One important limitation is that only simulated environmental sensors can be used in driving simulators [18]. As the environment and the surrounding traffic are virtual, these data are used as input for the system algorithms. Experiments regarding the sensor performance itself can thus not be conducted in a driving simulator. This can be an advantage, if no highly developed sensors are available or if any uncertainty due to the (imperfect) sensing equipment should not distort the results of the experiment.

However, several important points have to be considered if driving simulator experiments are conducted or the results interpreted. Depending on the technological level of the simulator, an experiment can be very complex and can end up at high costs [80]. Many simulators, especially the more advanced ones including dynamics simulation, are immovable and thus result in geographic inflexibility of the experiment, which can also influence the structure of the subject sample [80]. Even highly advanced dynamic simulators have limited abilities concerning realistic driving dynamics [79]. On the contrary, motion simulation comes at the possible price of motion sickness [79, 80], which results in loss of data for a fraction of the sample affected. As the environment, and so to say the "world", the subject is in are virtual, specific motivational aspects relevant for driver behavior can be distorted; consequently behavior can differ from that of real traffic [18]. The sometimes "clinical" look and feel of situations can also lead to a lack of perception regarding criticality and can produce other behavioral artifacts [80]. Therefore, the validity of the simulation should be proven for every research question [49]. Driving simulator experiments are always restricted to a limited number of situations [18]. As a consequence, overall effects of a measure cannot be assessed in driving simulators [65],

Fig. 2.3 Vehicle in the Loop: Vehicle, head-mounted display, and head-tracking [79]

and experiments can hardly be regarded as representative, for example, in the sense of overall effectiveness in a traffic system [2].

The limitations imposed on subject studies by test track as well as driving simulators have inspired a new approach. The idea of *Vehicle in the Loop* (ViL) is a combination of the advantages of track testing with driving simulators while avoiding some of their limitations [49, 83, 88, 89]. The basis is a real vehicle in combination with a virtual environment (see Fig. 2.3). The vehicle drives on a test track, but critical objects in the environment (e.g., other traffic participants) are virtual. The information for the vehicle system under investigation thus comes from the virtual environment, but triggers real system responses within the real vehicle. The subject wears an optical head-mounted display. Virtual objects are projected in an appropriate way into the real spatial environment according to the contact analogue paradigm. In the *augmented* mode, *some* virtual objects are projected into the real environment. In the *virtual* mode, *everything* the subject sees is virtual [49, 83].

The striking advantage of ViL is the real vehicle including obviously realistic movement and vehicle responses. The experiments are nevertheless safe, even in subjectively critical situations. The reproducibility is high. The method has its strongest advantages in safety critical situations when realistic driving dynamics are of importance [49]. Reported drawbacks are dimension and weight of the head mounted display, which can result in changed driver behavior and headache, whereas motion sickness has not been observed [49]. As in every method, validity is of high importance. Several aspects of driver behavior have been investigated and compared to responses in "normal" vehicles on a test track. For example, the following distance to lead vehicles, several reaction times, and general driving patterns were found to be similar. Acceleration in curves and recognition of the lane of distant vehicles were found to be not exactly comparable [49].

Another possibility for the testing of both driver as well as system behavior are studies in *real traffic*. Obviously, the whole surrounding is realistic and thus provides a maximum of validity in this respect [80]. Testing is geographically very flexible and allows an investigation of "normal" driving behavior under various circumstances. It is also possible to test false-positive system reactions, triggered by a variety of (random) influences [80]. This can be done with a deactivated system (i.e., open-loop), which means the system works in the background and its output is recorded, but no interaction with the driver or vehicle is allowed [9]. This kind of testing can only be carried out rather late during development [9], as functions and components must have approval for testing in traffic. Depending on the function in question, additional safety measures must be included to ensure safe testing [80]. One main deficit of testing in real traffic is that specific conditions or scenarios can hardly be triggered and cannot be reproduced easily [49, 80]. A systematic variation of conditions, such as in driving simulators, requires a high effort [80]. The true-positive reaction of safety systems cannot be tested at all, as testing must always be safe for every participant involved [49].

There are several techniques for the analysis of driving behavior in its natural environment, i.e., real traffic [44]. They are summarized under the term *Naturalistic Driving Study* (NDS). An NDS is "[...] the observation of drivers in naturalistic settings (during their regular, everyday driving) in an unobtrusive way. The essential driver behavior is what is of interest in these studies, usually in relation to crashes" [44]. If a system is included in the observation, it is called *Field Operational Test* (FOT). An FOT can include quasi experimental methods and is focused on behavior in combination with a system in the field [44]. Examples for NDS are the 100-Car Naturalistic Driving Study [90] and The Second Strategic Highway Research Program (SHRP2) [91]. Some examples for FOTs are euroFOT [92] or the Integrated Vehicle-Based Safety Systems: Light-Vehicle Field Operational Test (IVBSS) [93].

The consideration applying to testing in real traffic with regard to subjects also apply to NDS and FOT. The advantages of these observational methods in real traffic are that they provide the only way of discovering unexpected behavioral patterns, especially in combination with a safety system [44]. Over an extended observational interval, they provide a very reliable source of information on driver behavior [44] and also generate knowledge on traffic and environment. One crucial point is that these studies allow for an estimation of exposure in various forms, which is not feasible in the methods described above [43]. The downsides are that experimental control is extremely limited, that different methods have been applied in nearly all studies conducted so far, and that these studies require extremely high efforts and costs [44]. Although these methods are the only ones presented in this section which are capable of capturing "real-world effectiveness", the information derived is in the context of this thesis rather used to derive models.

2.8 Summary and Conclusion

This chapter has summarized the state of scientific and technical knowledge concerning the evaluation of the pre-crash phase. With respect to the fundamentals of traffic as given in Chap. 1, basic methodological aspects of evaluation have been discussed. The difference between *accident-* and *traffic-*based testing and the subsequent meaning of possible findings are discussed. Validity of findings is an overall issue that is of high importance regardless of the method used.

The evaluation of safety functions, especially of active and integral safety, is not limited to true-positive system responses. The second section has elaborated on all possible system responses, classified in true positive (or negative) and false positive (or negative). The importance of these terms for system development regarding the operating point as well as system optimization and the implications for evaluation are discussed.

The review of existing methods of evaluation includes a general description of retrospective and prospective testing and data sources available. For new systems of active safety a prospective approach seems nearly always to be most promising. However, the data sources available have deficits regarding the amount of data as well as depth of information and quality. Different methods of evaluation have been introduced; especially their advantages and challenges with respect to validity have been discussed.

Given the objective of prospective evaluation regarding the overall effectiveness of a measure of active safety representative for a traffic system, the methods discussed above—ranging from analysis of accident data bases to sophisticated case-by-case simulations—do not seem to be adequate. Mass simulations covering all relevant varieties (e.g., due to human behavior) as well as uncertainties in different situations offer an approach to meet this objective.

As mass simulation seems to be the method of choice for representative evaluation of the effectiveness of systems of active and integral safety, substantial modeling and input data are necessary. The last section has introduced different methods used to gain knowledge needed for implementation in such a simulation. Starting with methods focused on single components or subsystems, examples of different techniques and expected results have been given. The next level is the testing of whole systems, for example, on test tracks or in subject studies. Research on driver behavior is also a vital part, often conducted in driving simulators, on test tracks, or in real traffic. Exposure and long term studies can be conducted as FOT or NDS in real traffic and provide valuable input for modeling different parts of the driver, vehicle and environment system in a simulation or enable validation or even verification of different parts of a process model, especially for critical situations.

References

1. Green, M. (2000). "How long does it take to stop?" methodological analysis of driver perception-brake times. *Transportation Human Factors, 2*(3), 195–216.
2. Gründl, M. (2005). *Fehler und Fehlverhalten als Ursache von Verkehrsunfällen und Konsequenzen für das Unfallvermeidungspotential und die Gestaltung von Fahrerassistenzsystemen.* Dissertation, Universität Regensburg.
3. Kleinbaum, D., & Klein, M. (2010). *Logistic Regression. A Self Learning Text.*, Statistics for Biology and Health Berlin: Springer.
4. Kates, R., Jung, O., Helmer, T., Ebner, A., Gruber, C., & Kompass, K. (2010). Stochastic simulation of critical traffic situations for the evaluation of preventive pedestrian protection systems. In *Erprobung und Simulation in der Fahrzeugentwicklung.*
5. Cook, R. J., & Sackett, D. L. (1995). The number needed to treat: A clinically useful measure of treatment effect. *British Medical Journal, 18*, 452–454.
6. Schechtman, E. (2002). Odds ratio, relative risk, absolute risk reduction, and the number needed to treat—Which of these should we use? *Value in Health, 5*(5), 431–436.
7. Rosen, E., Källhammer, J.-E., Eriksson, D., Nentwich, M., Frederiksson, R., & Smith, K. (2009). Pedestrian injury mitigation by autonomous braking. In *21st International Technical Conference on the Enhanced Safety of Vehicles (ESV 2009)*, No. 09–0132.
8. Wisselmann, D., Gresser, K., Hopstock, M., & Huber, W. (2009). Präventiver statt passiver Fußgängerschutz. In *AAET 2009 - Automatisierungssysteme, Assistenzsysteme und eingebettete Systeme für Transportmittel* (pp. 60–76). Gesamtzentrum für Verkehr Braunschweig e.V.
9. Fach, M., & Ockel, D. (2009). Evaluation methods for the effectiveness of active safety systems with respect to real world accident analysis. In *21st International Technical Conference on the Enhanced Safety of Vehicles (ESV 2009)*, No. 09–0311.
10. Van Auken, R. M., Zellner, J. W., Chiang, D. P., Kelly, J., Silberling, J. Y., Dai, R., et al. (2011). Advanced Crash Avoidance Technologies (ACAT) Program—Final Report of the Honda-DRI Team, Volume I: Executive Summary and Technical Report. Technical Report DOT HS 811 454A, Dynamic Research Inc./National Highway Traffic Safety Administration.
11. Aoki, H., Aga, M., Miichi, Y., & Matsuo, Y. (2009). Development of a safety impact estimation tool for advanced safety technologies. In *21st International Technical Conference on the Enhanced Safety of Vehicles (ESV 2009)*, No. 09–0025.
12. Bakker, J., & Herrmann, R. (2001). Systematik zur Ermittlung der Wirksamkeit von Fahrzeug-Sicherheitsmaßnahmen auf der Grundlage von deutschen Unfalldatenbanken. In *Innovativer Kfz-Insassen- und Partnerschutz: Tagung Berlin, 6. - 7. September 2001* (Düsseldorf, 2001), No. 1637 in VDI-Berichte, VDI Verlag.
13. Busch, S. (2005). Entwicklung einer Bewertungsmethodik zur Prognose des Sicherheitsgewinns ausgewählter Fahrerassistenzsysteme. In *Fortschritts-Berichte VDI*, No. 588 in 12. VDI Verlag, Düsseldorf, 2005.
14. Hannawald, L. (2008). *Multivariate Bewertung zukünftiger Fahrzeugsicherheit.* Dissertation, Technische Universität Dresden.
15. Schramm, S., & Roth, F. (2009). Method to assess the effectiveness of active pedestrian protection systems. In *21st International Technical Conference on the Enhanced Safety of Vehicles (ESV 2009)*, No. 09–0398.
16. Kreiss, J.-P., Stanzel, M., & Zobel, R. (2011). On the use of real-world accident data for assessing the effectiveness of automotive safety features—Methodology, timeline and reliability. In *22st International Technical Conference on the Enhanced Safety of Vehicles (ESV 2011)*, no. 11–0054.
17. Breuer, J. (2002). Objektive Bewertung der Aktiven Sicherheit von Fahrzeugen: Kritische Betrachtung. In *VDA Technischer Kongreß 2002 - Sicherheit durch Elektronik* (Frankfurt a. Main, 2002) (pp. 105–113). VDA - Verband der Automobilindustrie.
18. Fastenmeier, W., Gstalter, H., & Zahn, P. (2001). Prospektive Risikopotentialabschätzung am Beispiel der Spurwechsel-Assistenz. In *Der Fahrer im 21. Jahrhundert* (Düsseldorf, 2001), No. 1613 in VDI-Berichte, VDI Verlag.

19. Zangmeister, T., Kreiß, J.-P., Page, Y., & Cuny, S. (2009). Evaluation of the safety benefits of passive and/or on-board active safety applications with mass accident data-bases. In *21st International Technical Conference on the Enhanced Safety of Vehicles (ESV 2009)*, No. 09–0222.

20. Schäbe, H., & Schierge, F. (2007). Investigation on the influence of car lighting on nighttime accidents in germany. In *7th International Symposium on Automotive Lighting—ISAL 2007—Proceedings of the Conference*, Technische Universität Darmstadt, München: Herbert Utz Verlag GmbH. ISBN-13: 978-3-8316-0711-2.

21. Schramm, S. (2011). *Methode zur Berechnung der Feldeffektivität integraler Fußgängerschutzsysteme*. Dissertation, Technische Universität München.

22. Statistisches Bundesamt. (2011). Verkehr - Verkehrsunfälle - 2010. No. Fachserie 8 Reihe 7. Statistisches Bundesamt, Wiesbaden.

23. Kuehn, M., Hummel, T., & Bende, J. (2009). Benefit estimation of advanced driver assistance systems for cars derived from real-life scenarios. In *21st International Technical Conference on the Enhanced Safety of Vehicles (ESV 2009)*, No. 09–0317.

24. Kuehn, M., Hummel, T., & Bende, J. (2011). Advanced driver assistance systems for trucks—Benefit estimation from real-life accidents. In *22st International Technical Conference on the Enhanced Safety of Vehicles (ESV 2011)*, no. 11–0153.

25. GIDAS German In-Depth Accident Study. http://www.gidas.org/files/GIDAS_eng.pdf, as of December 22, 2010.

26. Otte, D., Haasper, C., & Wiese, B. (2008). Straßenwirksamkeit von Fahrradhelmen bei Verkehrsunfällen von Radfahrern auf Kopfverletzungshäufigkeit und Verletzungsschwere. Verkehrsunfall und Fahrzeugtechnik (VKU) (Oktober 2008), pp. 2–12.

27. Grömping, U., Pfeiffer, M., & Stock, W. (2007). Deliverable 7.1 Statistical Methods for Improving the Usability of Existing Accident Databases. Technical Report. Project No. 027763—TRACE.

28. Hautzinger, H., Pfeifer, M., & Schmidt, J. (2006). Hochrechnung von Daten aus Erhebungen am Unfallort. BASt-Bericht F 59, Institut für angewandte Verkehrs- und Tourismusforschung e.V.

29. UMTRI. (2005). 1994–1998 NASS Pedestrian Crash Data Study (PCDS) Codebook. Version 03Mar01. UMTRI Transportation Data Center.

30. Hannawald, L., & Kauer, F. (2004). *Equal Effectiveness Study on Pedestrian Protection*. Dresden: Technische Universität Dresden.

31. Schijndel de Nooij, M. V., Hair-Buijssen, S. D., Versmissen, T., Fredriksson, R., Rosen, E., & Olsson, J. (2011). Holland: VRU paradise goes for the next safety level. In *22st International Technical Conference on the Enhanced Safety of Vehicles (ESV 2011)*, No. 11–0094.

32. Reichart, G. (2002). Vom Fehler zum Unfall - Ein neuer Ansatz in der Unfallforschung. In *VDA Technischer Kongreß 2002 - Sicherheit durch Elektronik* (pp. 59–76), VDA - Verband der Automobilindustrie, Frankfurt/Main: VDA Verband der Automobilindustrie.

33. Hakkert, A. S., Gitelman, V., & Vis, M. A. (2007). Road Safety Performance Indicators: Theory. Deliverable D3.6, EU FP6 project SafetyNet.

34. Kocherscheidt, H. (1993). Möglichkeiten und Grenzen einer Fahrzeugsicherheitsbewertung. In *Innovativer Kfz-Insassen- und Partnerschutz*, no. (1046) in VDI-Berichte. Düsseldorf: VDI Verlag.

35. Gstalter, H. (1983). *Der Verkehrskonflikt als Kenngröße zur Beurteilung von Verkehrsabläufen und Verkehrsanlagen*. Dissertation, Technische Universität Carolo-Wilhelmina Braunschweig.

36. Schlag, B., & Weller, G. (2002). Kriterien zur Beurteilung von Fahrerassistenzsystemen. In *38. BDP-Kongress für Verkehrspsychologie Universität Regensburg 2002*. Bonn: Berufsverband Deutscher Psychologinnen und Psychologen (BDP).

37. Fastenmeier, W., & Gstalter, H. (2000). Risikobewertung des Straßenverkehrs in definierten Verkehrssituationen/Ermittlung eines "Gefährlichkeitsindices". Projektskizze/Mögliche Arbeiten.

38. Der Finkelstein, M. M. (1990). Der zukünftige Bedarf der Verkehrssicherheitsforschung liegt beim Thema aktive Sicherheit. *Zeitschrift für Verkehrssicherheit, 36*(4), 155–158.

39. Zimmermann, M., Georgi, A., Lich, T., & Marchthaler, R. (2009). Nutzenanalyse für Auf-
fahrunfälle vermeidende Sicherheitssysteme. In *AAET 2009 - Automatisierungssysteme, Assistenzsysteme und eingebettete Systeme für Transportmittel* (pp. 78–86), Gesamtzentrum für
Verkehr Braunschweig e.V., Ed.
40. Gottselig, B., Eis, V., Wey, T., & Sferco, R. (2008). Entwicklung der Verkehrssicherheit -
Potential-Bestimmung von modernen Sicherheitssystemen im "Integralen Ansatz". In *VDA 10.
Technischer Kongress (2008)* (pp. 173–184). VDA - Verband der Automobilindustrie, Frank-
furt/Main: VDA Verband der Automobilindustrie.
41. Scholliers, J., Blosseville, J. M., Netto, M., Mammar, S., Chen, J., Heinig, K., et al. (2007).
Review of validation procedures for preventive and active safety functions. PReVENT—
Preventive and Active Safety Applications Integrated Project—Contract number FP6-
507075—SP Deliverable D16.1.
42. Donges, E. (1999). A conceptual framework for active safety in road traffic. *Vehicle System
Dynamics, 32*, 113–128.
43. Burgett, A., Srinivasan, G., & Ranganathan, R. (2008). A methodology for estimating potential
safety benefits for pre-production driver assistance systems. Final Report DOT HS 810 945,
National Highway Traffic Safety Administration.
44. Lietz, H., Petzold, T., Henning, M., Haupt, J., Wanielik, G., Krems, J., et al. (2008). Methodische
und technische Aspekte einer Naturalistic Driving Study. FAT Schriftenreihe 229 BASt FE
82.0351/2008, Forschungsvereinigung Automobiltechnik e.V. (FAT).
45. Reichart, G. (2001). *Menschliche Zuverlässigkeit beim Führen von Kraftfahrzeugen.* No. 7 in
22 Mensch-Maschine-Systeme. Düsseldorf: VDI Verlag.
46. Nirschl, G., Böttcher, S., Schlag, B., & Weller, G. (2004). Verfahren zur Bewertung der
Verkehrssicherheit von Fahrerassistenzsystemen durch objektive Erfassung von Fahrfehler-
risiken. In *Integrierte Sicherheit und Fahrerassistenzsysteme* (Vol. 1864, pp. 397–420). Düsseldorf: VDI Verlag.
47. Digges, K. H., Nicholson, R. M., & Rouse, E. J. (1985). The Technical Basis for the Center
High Mounted Stoplamp. *SAE Technical Paper*, 851240.
48. Bours, R., & Tideman, M. (2010). Simulation tools for integrated safety design. In *Proceedings
of Airbag 2010, International Symposium and Exhibition on Sophisticated Car Occupant Safety
Systems*.
49. Bock, T. (2008). *Vehicle in the loop: Test- und Simulationsumgebung für Fahrerassistenzsysteme.* Dissertation, Technische Universität München.
50. Bamberg, R., & Zellmer, H. (1994). *Nutzen durch fahrzeugseitigen Fußgängerschutz.* No.
F 5 in Berichte der Bundesanstalt für Straßenwesen, Fahrzeugtechnik. Bergisch Gladbach:
Bundesanstalt für Straßenwesen.
51. Liers, H., & Hannawald, L. (2011). Benefit estimation of secondary safety measures in real-
world pedestrian accidents. In *22st International Technical Conference on the Enhanced Safety
of Vehicles (ESV 2011)*, No. 11–0300.
52. Liers, H. (2010). Extension of the Euro NCAP effectiveness study with a focus on MAIS3+
injured pedestrians. Final report, VUFO GmbH.
53. Liers, H., & Hannawald, L. (2009). Benefit estimation of the EuroNCAP pedestrian rating
concerning real-world pedestrian safety. Verkehrsunfallforschung an der TU Dresden GmbH.
54. Reßle, A., Schramm, S., & Kölzow, T. (2010). Generierung von Verletzungsrisikofunktionen
für Fußgängerkollisionen. In *Crash Tech 2010 - Fahrzeugsicherheit 2020*.
55. Erbsmehl, C. (2009). Simulation of real crashes as a method for estimating the potential benefits
of advanced safety technologies. In *21st International Technical Conference on the Enhanced
Safety of Vehicles (ESV 2009)*, No. 09–0162.
56. Schubert, A., Erbsmehl, C., & Hannawald, L. (2013). Standardized pre-crash-scenarios in
digital format on the basis of the vufo simulation. In *5th International Conference on ESAR
"Expert Symposium on Accident Research"*, No. F 87 in Fahrzeugtechnik, Bundesanstalt für
Straßenwesen, Fachverlag nw.
57. Döring, S., Jungbluth, A., Labenski, V., & Wille, J. (2012). Effektivitätsbewertung mit rateEF-
FECT am Beispiel des vorausschauenden Fußgängerschutzes. In *Grazer SafetyUpDate 2012*.

58. Kohsiek, A., Zatloukal, M., & Wille, J. M. (2011). Entwicklung eines Werkzeugs zur Effizienzbewertung aktiver Sicherheitssysteme. In *Grazer SafetyUpDate 2011*.
59. Wille, J. M., Jungbluth, A., & Kohsiek, A. Zatloukal, M. (2012). rateEFFECT - Entwicklung eines Werkzeugs zur Effizienzbewertung aktiver Sicherheitssysteme. In *AAET 20012 - Automatisierungs-, Assistenzsysteme und eingebettete Systeme für Transportmittel*. Gesamtzentrum für Verkehr Braunschweig e.V.
60. Wille, J. M., Jungbluth, A., & Kohsiek, A. Zatloukal, M. (2012). rateEFFECT - Entwicklung eines Werkzeugs zur Effizienzbewertung aktiver Sicherheitssysteme. 5. Tagung Fahrerassistenz.
61. Wille, J. M., & Zatloukal, M. (2013). rateEFFECT effectiveness evaluation of active safety systems. In *5th International Conference on ESAR "Expert Symposium on Accident Research"*, No. F 87 in Fahrzeugtechnik, Bundesanstalt für Straßenwesen, Fachverlag nw.
62. Barrios, M. J., Aparicio, A., Davila, A., Miguel, J. L. D., Modrego, S., Olona, A., et al. (2009). Evaluation of the effectiveness of pedestrian protection systems through in-depth accident investigation, reconstruction and simulation. In *21st International Technical Conference on the Enhanced Safety of Vehicles (ESV 2009)*, No. 09–0376.
63. Hannawald, L., Erbsmehl, C., & Liers, H. (2011). Benefit assessment of forward-looking safety systems. In *22st International Technical Conference on the Enhanced Safety of Vehicles (ESV 2011)*, No. 11–0212.
64. Leneman, F., Verburg, D., & Hair-Buijssen, S. D. (2008). PreScan, testing and developing active safety applications through simulation. In *3. Tagung: Aktive Sicherheit durch Fahrerassistenz*.
65. Kusano, K. D., & Gabler, H. C. (2011). Potential effectiveness of integrated forward collision warning, per-collision brake assist, and automated pre-collision braking systems in real-world, rear-end collisions. In *22st International Technical Conference on the Enhanced Safety of Vehicles (ESV 2011)*, No. 11–0364.
66. Gordon, T., Sardar, H., Blower, D., Ljung Aust, M., Bareket, Z., Barnes, M., et al. (2010). Advanced Crash Avoidance Technologies (ACAT) Program—Final Report of the Volvo-Ford-UMTRI Project: Safety Impact Methodology for Lane Departure Warning—Method Development and Estimation of Benefits. Final Report DOT HS 811 405, USDOT / National Highway Traffic Safety Administration.
67. Harding, J. (2009). The advanced crash avoidance program (ACAT). In *Proceedings of the 16th ITS World Congress*.
68. Auken, V. R., Zeller, J., Silberling, J., Sugimoto, Y., & Urai, Y. (2011). Progress report on evaluation of a pre-production head-on crash avoidance assist system using an extended "safety impact methodology" (SIM). In *22st International Technical Conference on the Enhanced Safety of Vehicles (ESV 2011)*, No. 11–0207.
69. Euro NCAP. (2009). European New Car Assessment Programme (Euro NCAP) Assessment Protocol—Pedestrian Protection. No. Version 5.0. www.euroncap.com.
70. Euro NCAP. (2010). European New Car Assessment Programme (Euro NCAP) Pedestrian Testing Protocol. No. V 5.1. www.euroncap.com.
71. Anata, K., Konosu, A., & Issiki, T. (2011). Injury risk assessment at the timing of a pedestrian impact with a road surface in a car-pedestrian accident. In *22st International Technical Conference on the Enhanced Safety of Vehicles (ESV 2011)*, No. 11–0119.
72. Fredriksson, R. (2011). Priorities and Potential of Pedestrian Protection—Accident data, Experimental tests and Numerical Simulations of Car-to-Pedestrian Impacts. PhD thesis, Karolinska Institutet.
73. Fröming, R. (2008). Assessment of integrated pedestrian protection systems. In *Fortschritt-Berichte VDI*, No. 681 in 12. VDI Verlag GmbH, Düsseldorf.
74. Kühn, M., Fröming, R., & Schindler, V. (2005). *Fußgängerschutz - Unfallgeschehen, Fahrzeuggestaltung, Testverfahren*. Berlin: Springer.
75. Hamacher, M., Eckstein, L., Kühn, M., & Hummel, T. (2011). Assessment of active and passive technical measures for pedestrian protection at the vehicle front. In *22st International Technical Conference on the Enhanced Safety of Vehicles (ESV 2011)*, No. 11–0057.

76. http://www.tudelft.nl/fileadmin/UD/MenC/Support/Internet/TU_Website/TU_Delft_portal/ Actueel/Nieuwsberichten/Persberichten_-_ARCHIEF/2007/img/bvof-TNO_Automotive_1. jpg, as of March 26, 2013.

77. Gietelink, O., Ploeg, J., Schutter de, B., & Verhaegen, M. (2004). Testing advanced driver assistance systems for fault management with the VEHIL test facility. In *Proceedings of the 7th International Symposium on Advanced Vehicle Control (AVEC'04)* (pp. 579–584).

78. Kusters, L. J. J., Gietelink, O. J., Hoof, J. V., & Lemmen, P. P. M. (2004). Evaluation of advanced driver assistance system with the VEHIL test facility. In *Integrierte Sicherheit und Fahrerassistenzsysteme*, No. 1864 (pp. 421–437). Düsseldorf: VDI Verlag.

79. Bock, T. (2009). Bewertung von Fahrerassistenzsystemen mittels der Vehicle in the Loop-Simulation. In H. Winner, S. Hakuli, & G. Wolf (Eds.), *Handbuch Fahrerassistenzsysteme* (pp. 76–83). Wiesbaden: Vieweg+Teubner Verlag/GWV Fachverlage GmbH.

80. Breuer, J. (2009). Bewertungsverfahren von Fahrerassistenzsystemen. In H. Winner, S. Hakuli, & G. Wolf (Eds.), *Handbuch Fahrerassistenzsysteme* (pp. 55–68). Wiesbaden: Vieweg+Teubner Verlag/GWV Fachverlage GmbH.

81. Fecher, N., Regh, F., Habenicht, S., Hoffmann, J., & Winner, H. (2008). Test- und Bewertungsmethoden für Sicherheitssysteme der Bahnführungsebene. *Automatisierungstechnik*, *11*(56), 592–600.

82. Hoffmann, J., & Winner, H. (2008). EVITA - Das Untersuchungswerkzeug für Gefahrensituationen. In *3. Tagung: Aktive Sicherheit durch Fahrerassistenz*.

83. Hoffmann, J., & Winner, H. (2009). EVITA - Das Prüfverfahren zur Beurteilung von Antikollisionssystemen. In H. Winner, S. Hakuli, & G. Wolf (Eds.), *Handbuch Fahrerassistenzsysteme* (pp. 69–75). Wiesbaden: Vieweg+Teubner Verlag/GWV Fachverlage GmbH.

84. Jandl, C. (2011). Entwicklung eines Crashobjektes zur schadlosen Kollision mit Kraftfahrzeugen für die Erprobung aktiver Sicherheitssysteme. Diplomarbeit, FH Johanneum.

85. Roth, F., Stoll, J., Zander, A., Schramm, S., & von Neumann-Cosel, K. (2008). Methodik zur Funktionsentwicklung des vorausschauenden Fußgängerschutzes. In *24. VDI / VW-Gemeinschaftstagung - Integrierte Sicherheit und Fahrerassistenzsysteme* (Düsseldorf, 2008) (Vol. 2048, pp. 177–190). VDI-Berichte, VDI Verlag

86. Hoffmann, J. (2008). *Das Darmstädter Verfahren (EVITA) zum Testen und Bewerten von Frontalkollisionsgegenmaßnahmen*. Dissertation, Technische Universität Darmstadt.

87. Carsten, O., & Nilsson, L. (2001). Safety assessment of driver assistance systems. *European Journal of Transport and Infrastructure Research, 1*, 225–243.

88. Bock, T., Maurer, M., & Färber, G. (2007). Validation of the vehicle in the loop (VIL)—A milestone for the simulation of driver assistance systems. In *Proceedings of the 2007 IEEE Intelligent Vehicles Symposium* (pp. 612–617).

89. Bock, T., Maurer, M., & van Meel, F. (2008). Vehicle in the Loop: Ein innovativer Ansatz zur Kopplung virtueller mit realer Erprobung. *ATZ - Automobiltechnische Zeitschrift, 110*(1), 2–8.

90. Dingus, T. A., Klauer, S. G., Neale, V. L., Petersen, A., Lee, S. E., Sudweeks, J., et al. (2006). The 100-Car Naturalistic Driving Study, Phase II—Results of the 100-Car Field Experiment. No. DOT HS 810 593. Department of Transportation, National Highway Traffic Safety Administration, Washington.

91. Antin, J., Lee, S., Hankey, J., & Dingus, T. (2011). Design of the in-vehicle driving behavior and crash risk study. In Support of the SHRP 2 Naturalistic Driving Study. Report SHRP2 S2–S05-RR-1, Virginia Tech Transportation Institute.

92. Benmimoun, M., & Benmimoun, A. (2010). Large-scale FOT for analyzing the impacts of advanced driver assistance systems. In *Proceedings of the 17th ITS World Congress*.

93. Sayer, J. R., Bogard, S. E., Buonarosa, M. L., LeBlanc, D. J., Funkhauser, D. S., Bao, S., et al. (2011). Integrated Vehicle-Based Safety Systems. Light-Vehicle Field Operational Test Key Findings Report. Report DOT HS 811 416, The University of Michigan: Transportation Research Institute.

Chapter 3
Approach to Integrated Safety Evaluation: Preventive Pedestrian Protection

3.1 Process Chain for Quantitative Evaluation of the Pre-crash Phase

The concept for the process chain for the evaluation of measures taken before a collision responds to the challenges in evaluating real-world safety benefits using methods as described in Chap. 2. The objectives and requirements of the process chain can be summarized as follows:

- The method should predict real-world safety benefits of measures applied in the pre-crash phase.
- The method should produce a quantitative and representative evaluation of real-world effectiveness.
- The method should be objective, reliable, valid, reasonable, economic, free of feedback, safe, and privacy protecting [1].
- Possible, yet undesired aspects of a measure (such as false-positive actions) should also be part of the evaluation in order to predict the overall effect on safety as well as other impacts such as acceptance or efficiency.

The effectiveness of new measures should be quantitatively evaluated during the design and development phases, i.e., before market introduction, so approaches using retrospective analysis (e.g., based on accident data) are not applicable. The method must therefore not only be valid in the sense that it is able to capture the desired effect, but also be valid in its structure, assumptions, and internal procedures in order to produce a realistic and meaningful result. Therefore, real-world effectiveness requires statistical representativity. Classical methods such as subject studies in driving simulators lack this representativity considering a combination of different possible variations (e.g., subject sample, environmental conditions, etc.). The method of choice to fulfill these requirements is a simulation technique.

A stochastic simulation can fulfill the requirements of representativity, economy, safety, and privacy. Reliability and validity of the procedure have to be evaluated. The method itself has evidently no feedback on the subject under investigation. As no

© Springer International Publishing Switzerland 2015
T. Helmer, *Development of a Methodology for the Evaluation of Active Safety using the Example of Preventive Pedestrian Protection*, Springer Theses,
DOI 10.1007/978-3-319-12889-4_3

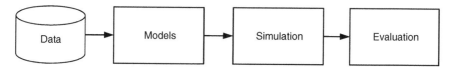

Fig. 3.1 Process chain for the evaluation of the pre-crash phase: overview

subjects are involved in the simulation, questions of ethics and reasonableness do not impose limitations. Stochastic simulation can also support evaluation of "undesired" system actions and their side effects. For an overall evaluation of safety effects, undesired system actions (i.e., false positives or false negatives) can reduce the safety benefit by not addressing relevant situations or in the worst case possibly provoking new hazardous situations induced by the system actions.

The general outline of the whole process is illustrated in Fig. 3.1. The basis for evaluation are data sources of various kinds as input for detailed modeling. Basically, a stochastic simulation generates virtual traffic including the vehicle with and without the measure in question as well as other participants, the relevant environmental and boundary conditions. The results are evaluated regarding positive and negative safety effects of the measure.

Figure 3.2 gives some details on the different steps of the process chain for the example of preventive pedestrian protection. Concerning *data* used, knowledge regarding the driver and pedestrian behavior (if not extractable from accident data) are taken from literature. The vehicle and preventive pedestrian protection related aspects are also based on literature as well as corporate knowledge. The intention is to construct evidence-based models using well-established statistical information to the greatest extent possible. The experiments and methods described in Chaps. 2 and 4 are intended to provide information necessary for developing the different models. In case specific parameters are unknown or for some reason cannot be investigated, sensitivity analyses are utilized to quantify the resulting uncertainties.

The *modeling* step contains data preparation and aggregation, as well as development and assessment of models. The first part of this step is the construction of reference scenarios (see Sect. 3.2). Another is an operationally defined model of the preventive system and its effects on the other participants in the traffic situation (see Sect. 3.3). The implementation of the driver, the pedestrian, and the vehicle into an appropriate traffic model including boundary conditions is briefly described in Sect. 3.4 together with the simulation itself.

The *simulation* provides the software environment in order to process the input data and correctly manage the interaction of the included models (see Sect. 3.4). For example, each single scenario could be simulated with and without the measure in question or the whole virtual scenario, including a high number of individual situations, could be simulated one time with and one time without the measure. All relevant characteristics are recorded for the evaluation step. The main advantage of this procedure is the possibility of a realistic consideration of all relevant distributions (e.g.,

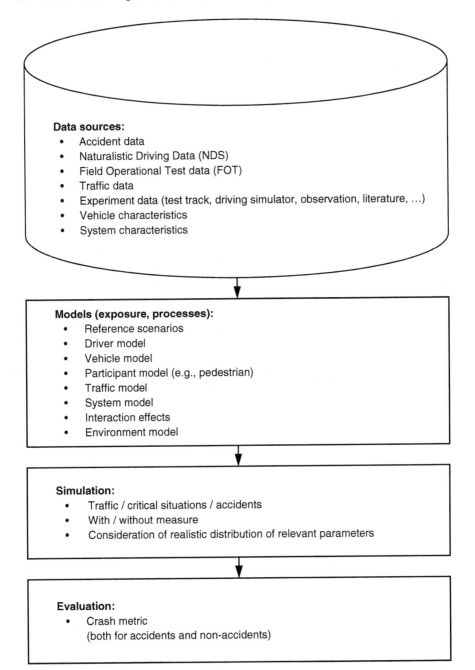

Fig. 3.2 Process chain for the evaluation of the pre-crash phase: details

driver reaction times, vehicle responses, sensor reliability, etc.). This characteristic gives the simulation the attribute *stochastic*.

The last step in the process chain is *evaluation*. The metric used consists of several steps assessing accident situations as well as non-accident situations on both a microscopic and a macroscopic level (see Sect. 3.5).

An important issue concerning the whole process chain is validation. An effective validation strategy usually begins at the level of a sub-process model. For example particular behavioral aspects of participants include perception of the pedestrian by the driver or the distinct driver reaction stages. Composite processes, such as time for the driver to respond to a complex stimulus, can be simulated by combining these sub-process models and can be independently validated. On the global level, composite processes are again combined to produce outcomes of interest and also secondary data, which are subject to validation as well, e.g., rate of accidents, influence of a change in an environmental condition. As explained in Chap. 2 (p. 17), verification may be possible for some subprocesses and processes, but hardly for all. The data used for validation and verification can come from accident statistics, secondary data sources such as other comparable studies, literature or experiments.

3.2 Reference Scenarios for Pedestrian Accidents

Accidents in general are unique events. Considering the near infinite variability of an accident in all its parameters, hardly any accident will happen twice the same way. Nevertheless, specific critical situations, which lead to accidents, as well as the accidents themselves, do show specific patterns and thus allow for a grouping. As a measure of traffic safety is aimed at addressing a particular group of situations and not one single accident, it is necessary to identify the criteria relevant for grouping and to characterize the distributions of key influencing parameters.

The individual accidents are grouped into so-called "reference scenarios". Reference scenarios are defined as "a limited number of scientifically derived traffic situations that represent a major part of the real traffic system" [2]. Basis for the construction of reference scenarios are in this case in-depth accident data. A detailed description of the data sources, the development of the methodology as well as the results can be found in [2–4]. The grouping uses parameters that have a high influence on the genesis of the critical situation and describe the pre-crash phase in a meaningful way. In the case of pedestrian accidents the key criteria are:

- The movement of the vehicle (e.g., moving straight or turning).
- The movement of the pedestrian (e.g., along the street or crossing).
- The site of the accident (i.e., urban, non-urban).

These criteria are used to derive general scenarios and sort them by frequency. The analysis using German accident data (namely GIDAS [5]) produced the following results (Table 3.1) for passenger vehicle-to-pedestrian accidents [4]. The results show that a vehicle going straight with a crossing pedestrian in an urban setting is by far

Table 3.1 Top six reference scenarios for pedestrian accidents, sorted by frequency [4]

No.	Scenario	Frequency (%)
1	Vehicle going straight, pedestrian crossing from the right, urban	43.5
2	Vehicle going straight, pedestrian crossing from the left, urban	28.2
3	Vehicle backing up, pedestrian crossing, urban and non-urban	10.0
4	Vehicle turning left, pedestrian crossing, urban	7.9
5	Vehicle turning right, pedestrian crossing, urban	2.7
6	Vehicle going straight, pedestrian moving straight, urban	2.4

the most frequent scenario (with about ¾ of all accidents). Taking the two turning vehicle scenarios (4, 5) together, they become more important than the backing up scenario (3). Scenarios with a pedestrian moving along the street are not very common, representing less than 2.5 % of all accidents.

The second dimension to determine the importance of reference scenarios besides frequency of occurrence is the severity of the accidents [2, 3], i.e., injury severity and property damage. Thus, in view of the personal dimension of suffering and tragedy resulting from severe injury, more severe accidents should be weighted more heavily. One way to attribute increasing weight to increasing injury severity is the HARM method [6, 7]. Each level of injury severity is expressed in monetary costs for society (including medical treatment, rehabilitation costs, reduction in productivity, etc.). As a consequence, more severe injuries are attributed with a higher factor [2, 3].

Comparing the importance of pedestrian reference scenarios using frequency and severity it becomes obvious that the overall picture given in Table 3.1 does not change much. The crossing scenarios (including 1 and 2) stay nearly unaltered, the turning scenarios become less important, and scenario 6 (pedestrian walking along the street) becomes more important. Data concerning the severity rating of backing up cases are not included in [2, 3]. The ranking of the scenarios does not change in this case due to the weighting by injury severity.

The results for the US show a comparable situation. The crossing scenarios with a straight moving vehicle are most important (by frequency and by severity), the turning scenarios are second by frequency and third by severity. The situation could be different in countries with left-hand driving. Backing up cases are less important or not included in the US data sets at all, which could be an effect of the sampling criteria applied during data collection rather than a representation of the accident situation [2, 3]. The same set of reference scenarios can therefore be used for Germany and the US regarding the criteria mentioned above.

The scenarios give important information about the critical situation and the constellation of the participants before the accident. In order to design measures of active safety, more detailed information about the scenarios is needed. Possible parameters of interest include, e.g., the time of day, initial speed of the vehicle, and collision speed of the vehicle. It is important to analyze every additional parameter not for all accidents, but per scenario individually. Regarding the time of day, it is daylight in 60–80% of the accidents, depending on the scenario. The initial speed as well as the collision speed are subject to more variation depending on the scenario. Detailed distributions of parameters as well as further information can be found in [2–4].

The reference scenarios define the boundary conditions for system development. The ranking of the scenarios provide a weighting criterion for evaluation of effectiveness within these scenarios. The overall safety effect of a measure can be calculated depending on the scenario and the specific condition a system is able to operate under (e.g., due to sensors used).

3.3 Functional Demonstrator of a Preventive Pedestrian Protection System

A functional demonstrator of a preventive pedestrian system is studied within this thesis. The functions of the vehicle-based system are best explained using the phases of an accident as given in Fig. 1.2 (p. 4). The preventive system is described by an operational model. An operational model describes the relevant functional behavior of system components independently of the precise engineering implementation. The general escalation strategy before a crash is the following:

- Information (included in the phase of normal driving).
- Warning (if a conflict or critical situation occurs).
- Automatic intervention (if an accident is highly probable and thus practically unavoidable).

A set of specific parameters can be found, for example, in Sect. 4.2 (p. 68).
The algorithm of the system has the following basic components:

- Detection of the situation.
- Interpretation and evaluation of the situation.
- Decision and action.

If a pedestrian is detected by the system, the physically possible trajectories of the vehicle and the pedestrian are calculated during interpretation and evaluation of the situation. The possible future trajectories for the vehicle (depending on current speed, maximum steering angle, friction, and acceleration capabilities) define a spatial region wherein the vehicle can reach every point within a given time span (different methods of calculation are explained in [8]).

The pedestrian's future movement is also predicted using a model. The model was developed in the "Aktive mobile Unfallvermeidung und Unfallfolgenminderung durch kooperative Erfassungs- und Trackingtechnologie" project (AMULETT) and focuses on physiological capabilities of the pedestrian regarding possible change in direction and acceleration depending on current speed [9–13]. For any given future point in time, the model estimates probabilities for the predicted position of the pedestrian. The algorithm of the system combines both probabilities for the future positions of pedestrian and vehicle and calculates a collision probability [8].

If a specific collision probability is exceeded, the system decides to act and gives the appropriate feedback. The decision of the system is rule-based. Bases are the prediction of the movements of the two participants and the predicted collision probability. The system can have three different categories of states:

1. Recognizable by the driver: pre-warning, warning or automatic braking.
2. Not recognizable by the driver: Reconfiguration of the brake assist.
3. No system action.

The rules for activation define states in category 1. States in the categories 2 and 3 are defined in combination with rules for non-activation. The following cases are possible for category 1:

- A pedestrian is detected *within* the zone directly in front of the vehicle (in relation to the vehicle trajectory): The corresponding action is initiated immediately, based on the current TTC.
- A pedestrian is detected *outside* the zone directly in front of the vehicle (in relation to the vehicle trajectory): The predicted collision probability is used for the decision.

The escalation strategy reflects the different levels of danger of the situations and allows for different activation thresholds in each case.

In order to have an effective system, which does not produce a high rate of undesired system actions, rules for non-activation must be included. Possible reasons for a deactivation of the system are:

- The driver performs an evasive maneuver (steering, braking, accelerating or any combination): The driver seems to respond to the situation and the warning functions of the system are suppressed (state 1). State 2 stays active in case the driver needs assistance in his emergency braking maneuver. The evasive maneuver is classified with limits regarding steering wheel velocity, steering angle, position of the accelerator pedal or acceleration of activation of the accelerator pedal. Braking by the driver does not automatically mean the driver is reacting to the emergency situation. If the braking is not sufficient to avoid the accident, the system supports the driver in braking by increasing the deceleration appropriately.
- The system only operates within a defined speed range.
- The system does suppress any actions for a given period of time after a preceding system action in order not to confuse the driver by multiple actions within a defined time span.

The design and implementation of a preventive system proves to be a challenge for the evaluation of safety benefits due to its internal complexity. The understanding of the basic functions and states of the system is thus necessary and must be included in the model, as they have a large influence on the overall effects.

3.4 Simulation of Vehicle-Pedestrian Interaction

The challenge of an objective, reliable, and representative evaluation of the pre-crash phase is facilitated by stochastic simulation [14–18]. Stochastic simulation methods (often referred to as Monte-Carlo techniques) are well-established, for example, in physics, engineering, biology, or chemical engineering. The simulation used in this thesis was developed at BMW Group as part of a development project and is partly described in [19–21]. The different steps of the process chain including modeling, simulation, and the data used are described in this section. To this end, the most important aspects are highlighted together with their literature related to the simulation. This description does not purport to be a full documentation of the simulation.

An important consideration for the process chain is that traffic accidents are statistically rare events and that each accident is a unique event (see Chap. 1). Detailed accident data are available in various databases and are very helpful in determining the potential of safety measures. However, critical traffic situations or near-accidents are not included in any representative databases. Field Operational Tests, like euroFOT [22], or Naturalistic Driving Studies, like the 100-Car Naturalistic Driving Study [23], collect data of such events, but often lack representativity for the traffic system as a whole. A possible solution can be provided, for example, by methods such as fault-tree analysis (see [24, 25]) and stochastic simulation. The simulation described in the following is a stochastic simulation, as many of the subprocesses involved include distributed parameters, which are hard to account for, e.g., in a fault-tree analysis.

The principle of the simulation is to decompose a complex problem, i.e., pedestrian-vehicle accidents, into processes and subprocesses, which can be understood and well modeled. The qualitative and quantitative modeling of the subprocesses is essential for the quality of the whole simulation. The microscopic models must be defined and connected with probability distributions for each possible state. The failures within the processes can be defined and analyzed. For example, one participant could fail to observe the other and consequently provoke a conflict. In order to evaluate a change in safety due to a measure of preventive pedestrian protection, the entire sequence of events leading to a possible pedestrian accident must be taken into consideration and thus be modeled. This has several advantages: A large variety of influences (e.g., impairment, changed environmental conditions) and system actions (e.g., information or automatic emergency braking) can be investigated regarding their impact on key processes (e.g., the response time of the driver, defined as time between first sight of the pedestrian and first activation of the brake pedal). The stochastic approach also allows for representative sampling, as all relevant parameters are considered with their distributions, and the sample size itself is basically just limited by calculation time. At the end of the simulation, the resulting key parameters, either of an accident or a non-accident (such as impact speed or speed reduction due to a system action), can be evaluated using a metric (see Sect. 3.5). The general structure as well as exemplary single models are described below.

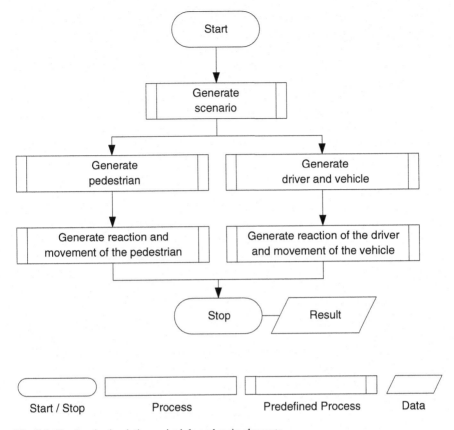

Fig. 3.3 Stochastic simulation: principle and main elements

Figure 3.3 gives the principle and the main elements of the stochastic simulation. The first step is the generation of the scenario, including the street and traffic. The next part is the generation of the pedestrian with his attributes. The driver and the vehicle with their attributes are generated in parallel. The vehicle includes the model for the measure of active safety under investigation. The simulation is time-based and evaluates each model and their interactions for each time step. In the end, either a collision happens or the pedestrian safely crosses the street.

All relevant processes are modeled and linked with realistic probability distributions. Each parameter is drawn randomly with respect to its probability distribution and possible dependencies on other factors in the simulation. The implemented scenario is an urban crossing scenario, as this is the most important one (see Sect. 3.2). The pedestrian crosses the street (straight road) from the right to the left from the view of the driver in the middle of a block. From the pedestrian's point of view, the traffic comes from the left. Scenario parameters include, for example, the geometry of the sidewalk, speed limit of the street or visibility restrictions. The traffic on the road itself is implemented as an exposure model depending on time of day and day of the week. The traffic state includes traffic volume, mean and standard deviation of speed, and also the gaps between the vehicles.

The pedestrian is described by age, gender, height, weight and alcohol level. The distribution of these attributes can itself depend on context variables, like time of day. The basic tasks to be fulfilled by the pedestrian in order to cross the street are displayed in Fig. 3.4. The pedestrian observes the traffic stream and decides when to cross. In this process, subprocesses for the perception of the distance and speed of the oncoming traffic as well as an estimation of the time required for crossing are included (these processes each are subject to human errors). Important is a model of gap acceptance, which compares the estimated crossing time with the estimated gap with respect to a model of impatience. The movement of the pedestrian, once he decided to cross the street, is determined by walking speed and crossing angle (again, each variable is distributed). If the pedestrian has made mistakes in his initial decision to cross the street (or severely underestimates the time needed to cross the street or overestimates TTC of the approaching vehicle), it is possible that he perceives a conflict. In that case, he estimates the acuteness of the situation and has an emergency response if he feels threatened. This can result in a change of speed and/or direction of movement.

Once the pedestrian has decided to cross the street and started into a specific gap of the traffic stream, a driver is modeled in the vehicle at the end of the gap. It can be that the gap is so large that no interaction at all will occur. In all other cases, the driver observes the situation (all cognitive processes are basically comparable for the pedestrian and the driver regarding their structure). The main steps in the driver model are given in Fig. 3.5. The perception process takes the central field of view into account and the probability for locating objects within the field of view. This is modeled as a change in angular size on the retina, which is a function of size of the object, distance, and relative speed. The reaction processes are modeled following the orient-observe-decide-act (OODA) paradigm as described in the literature [26–28]. Along the OODA process, the driver perceives the pedestrian and classifies the situation with respect to speed and position of the pedestrian (e.g., comfortable, emergency stopping, etc.). Braking is the only possible reaction of the driver implemented in the simulation considered here (emergency evasive maneuvers could be a different type of reaction). The efficacy of the braking itself is dependent on the time of braking, the intensity, the possible activation of a brake assistance system (which itself can be dependent on braking intensity), and the underlying physics of the combination of vehicle and road surface.

The vehicles in the simulation move on a straight street and have dimensions typical for mid-size vehicles. The braking capabilities are typical for up-to-date vehicles and typical road surfaces. The implemented preventive pedestrian protection system is thus modeled as part of the vehicle. Once the pedestrian is visible for the system, the probability per unit time that the pedestrian is detected by the system is modeled as a constant. The algorithm of the system includes a prediction of the vehicle's movement and the pedestrian's movement as well as the calculation of a collision probability as basis for a system action. The system itself has various stochastic components, e.g., inaccuracies regarding position and speed of the pedestrian. Depending

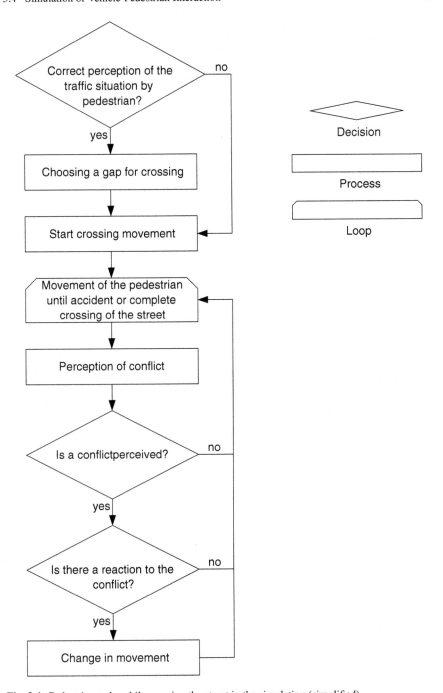

Fig. 3.4 Pedestrian tasks while crossing the street in the simulation (simplified)

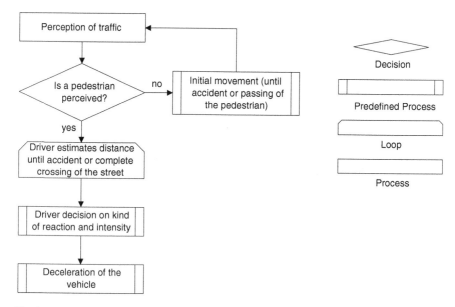

Fig. 3.5 Driver tasks while approaching the pedestrian in the simulation (simplified)

on the TTC calculated by the system, various actions as described in Sect. 3.3 can be triggered. The driver, if confronted with a warning, again performs the OODA loop, with respect to the specific distributions appropriate for the warning signal issued. These probability distributions depend on the current characteristics of the driver, the situation itself, and the design of the warning signal.

The simulation is terminated for each crossing once a collision has occurred or the pedestrian has safely crossed the street. In a particular scenario and modeling scheme, about one million crossings include about 2,200 collisions (without the preventive system activated). The reason for a rather high fraction of accidents is that the pedestrian of course can choose a gap in order to cross the street, but does not have a choice whether to cross at a safer location, e.g., crossing at an intersection. In addition to that, a model for impatience changes the pedestrian's gap acceptance with increasing waiting time. Thus, the scenario modeled has an increased inherent accident risk compared to safer crossing strategies.

Validity and plausibility of the simulation results are important aspects in the whole process. The plausibility of the various individual models was validated by considering the overall effects in comparison to accident data within the subsample of accidents (see Sect. 6.1, p. 143).

3.5 Evaluation of Safety Benefits

Evaluation of the safety benefits or—speaking more generally—the change in safety due to a specific measure is carried out after the simulation itself. Two parts make up the process of evaluation: methodology and metric. The metric in this case is closely linked to the method used for the simulation itself. Before giving examples of the dependencies between metric and methodology, a general explanation of the metric is given, which can be applied regardless of the simulation method.

Figure 3.6 gives different views and levels of an evaluation metric for active safety as well as exemplary indicators. First of all, there are two views: microscopic and macroscopic. The *microscopic* one focuses on a single event. This event can be an accident or also a non-accident (e.g., an avoided accident or a traffic situation in general). For every event there are three levels of evaluation:

- The first level is the *physical* level. An accident can be evaluated by comparing, e.g., the impact speed or the kinetic energy at the moment of collision. For a non-accident event, e.g., minimum distances or minimum TTC can used as safety indicators.
- The second level is the *physiological* level. It only exists in accidents and uses injury severity as metric. Several scales for injury severity are commonly used (see Sect. 5.1). The item of interest can be a single body part, a body region or the overall injury severity of the person.

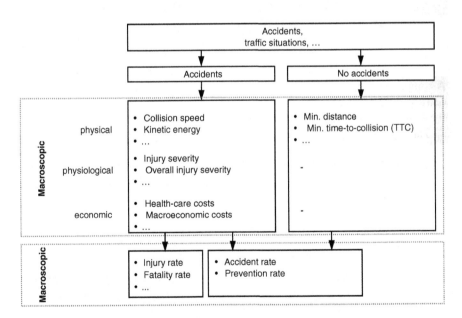

Fig. 3.6 Different views and levels of an evaluation metric for active safety

- The third level is the *economic* level. An accident can be evaluated in economic terms. An important distinction differentiates between medical costs (per single injury or per person) and societal costs, including medical treatment and rehabilitation costs, as well as the long-term reduction in productivity associated with the accident (see also HARM methodology, Sect. 3.2, p. 52).

The second view is *macroscopic*. In case more than one event is evaluated, an aggregation of the single events is possible in order to assess the overall effects. If the sample under investigation happens to contain accident and non-accident events, an accident rate or prevention rate can be calculated as ratio of frequency of accidents (or one minus accidents) *with* a measure by frequency of accidents *without* the measure. Summary statistics can also be computed in non-accident events by statistically evaluating the indicators defined on the physical level. In comparison to a baseline without measure the change due to a specific safety measure can be evaluated at the desired level of detail. Within the accident group, rates for specific injury severities as well as a fatality rate can be estimated.

The level of choice depends on the research question and the special interests of the researching institution. The physical level may be adequate and sufficient for, e.g., the comparison of different automatic emergency braking systems during the design and development phase. The physiological level is adequate if the protection potential of measures regarding injuries is under investigation. Questions related to insurance issues or subsidies could be more oriented toward an overall economic evaluation.

As mentioned above, the appropriate metric depends on the research question as well as on the methodology used in the experiment. Considering, for example, a hardware test with dummies, the metric has to be based on the readings of the dummy. Commonly used physical measurements such as the Head Injury Criterion (HIC) can then be translated into an injury probability, which is a well-known procedure [29]. A comparable approach is feasible using simulation. In a finite-element simulation or a kinematic simulation of collisions, a human model, e.g., the Total Human Model for Safety (THUMS), can be used [30].

The two different ways of measurement, i.e., dummy and virtual human model, provide physical data and utilize injury probability models to derive information about injury severities. Another approach is based on the change in injury level, using in-depth accident data and information about the injuries. The so-called Injury Shift Method evaluates the change in single injuries due to a particular measure [31]. Given a statistically significant amount of injury data in the data set, this method allows for a fast calculation of benefits, but compared to other methods mentioned relies on rather crude assumptions as elaborated above (see p. 31).

State of the art for evaluation on the physiological level are injury probability models (in case detailed collision simulations are not available or feasible). In many cases, e.g., if used in combination with a stochastic simulation, those models provide a translation of physical measurements at the moment of impact into physiological quantities. Considering pedestrians, models currently available are based on the person level (overall injury severity) and are mainly univariate using impact speed of the vehicle as explanatory variable [32–34] (see Chap. 5 for a detailed description and new multivariate approaches).

The third level, the economic evaluation, considers injuries as central part of the evaluation. Monetary costs can be attributed either on the level of single injuries [35], on the level of body regions [7], or overall for the whole person [36]. Some references [36, 37] give other factors related to injuries (e.g., medical, market productivity, legal costs) and non-injuries (e.g., travel delay, property damage).

There are several metrics at different levels for the evaluation of active safety. Depending on the research question and the method used, the quality of the assessment can vary. For example, injuries coded in an in-depth data base have a different reliability than injury probabilities given by probabilistic models or economic costs, which use injury information as basis. Stating the protection of the human as key objective, a metric based on injury severity seems to be appropriate.

3.6 Conclusion

This chapter discusses a process chain as solution to the challenge of predictive evaluation of active safety. Chapter 2 gave an overview of the current state of scientific and technical knowledge. The objectives as formulated at the beginning of this chapter prove to be hard to fulfill by currently available approaches. The approach presented here includes the modeling of a given situation in a traffic-based simulation using various data sources. The method of choice to achieve an objective and representative evaluation is a stochastic simulation connected with an appropriate metric.

The process chain starts with a reference scenario for the situation in question. For example, in the case of pedestrian protection, the most important scenario is a pedestrian crossing from the right in an urban setting (for Germany and the US). A functional demonstrator of a preventive pedestrian protection system is defined to test the process chain with a measure of active safety. The system detects the pedestrian, warns the driver, preconditions the brake assist, and as a last resort brakes automatically.

The traffic-based stochastic simulation of the crossing event used here has been developed in an internal project at BMW Group and was summarized here with respect to its structure and functions. The basic idea is a stochastic modeling of all processes from the pedestrian's decision to cross a road in the given scenario to the (avoided) accident. Each process is linked with appropriate probability distributions (mainly from literature). The pedestrian and the driver of the vehicle are implemented with respect to their individual attributes (e.g., age). The simulation results in uncritical crossings and in this specific scenario in about 0.2 % collisions. Overall effects in the accident events have been compared to data available in accident data bases to make a step towards validation of the model. The simulation described includes the whole process chain.

Besides having given an analysis of the state of the art in evaluating active safety and describing a possible solution to the problem, this thesis contributes to the investigation of driver behavior using an experiment in the dynamic driving simulator (see Chap. 4). The focus is also on the evaluation of the consequences of a vehicle-pedestrian collision: therefore a methodology for construction of injury probability models is developed (see Chap. 5). In the end, the results of the whole process chain are described and illustrated using a functional demonstrator of a preventive pedestrian protection system (see Chap. 6).

References

1. Winner, H., Hakuli, S., & Wolf, G. (Eds.). (2009). *Handbuch Fahrerassistenzsysteme*. Wiesbaden: Vieweg+Teubner Verlag/GWV Fachverlage GmbH.
2. Ebner, A., Samaha, R. R., Scullion, P., & Helmer, T. (2010). Methodology for the development and evaluation of active safety systems using reference scenarios: Application to preventive pedestrian safety. In *Proceedings of the International Research Council on Biomechanics of Injury (IRCOBI)* (pp. 155–168).
3. Ebner, A., Samaha, R. R., Scullion, P., & Helmer, T. (2010). Identifying and analyzing reference scenarios for the development and evaluation of preventive pedestrian safety systems. In *Proceedings of the 17th ITS World Congress*.
4. Helmer, T., Ebner, A., & Huber, W. (2009). *Präventiver Fußgängerschutz - Anforderungen und Bewertung*. Aachen: 18. Aachener Kolloquium Fahrzeug- und Motorentechnik.
5. GIDAS German In-Depth Accident Study. (2010). Retrieved December 22, 2010, from http://www.gidas.org/files/GIDAS_eng.pdf
6. Fildes, B., Gabler, H. C., Otte, D., Linder, A., & Sparke, L. (2004). Pedestrian impact priorities using real-world crash data and harm. In *Proceedings of the International Research Council on Biomechanics of Injury (IRCOBI)*.
7. Gabler, C., Digges, K., Fildes, B., & Sparke, L. (2005). Side impact injury risk for belted far side passenger vehicle occupants. *SAE Transactions, 114*(6), 34–42.
8. Kolbig, M. (2010). *Entwicklung eines Algorithmus für den präventiven Fußgängerschutz*. Leipzig: Diplomarbeit, Hochschule für Technik, Wirtschaft und Kultur Leipzig.
9. Projekt AMULETT. (2010). Retrieved August 15, 2010, from http://www.projekt-amulett.de
10. Meinecke, M., Roehder, M., Nguyen, T., Obojski, M., Heuer, M., Giesler, B., & Michaelis, B. (2009). Motion model estimation for pedestrians in street-crossing scenarios. In *7th International Workshop on Intelligent Transportation*.
11. Rasshofer, R. H., Schwarz, D., Morhart, C., & Biebl, E. (2009). Cooperative sensor technology for preventive vulnerable road user protection. In *21st International Technical Conference on the Enhanced Safety of Vehicles (ESV 2009)*. No. 09–0136.
12. Schmid, A. (2009). *Situationsanalyse und Kollisionswahrscheinlichkeitsberechnung für den präventiven Fußgängerschutz*. München: Diplomarbeit: Technische Universität München.
13. Zecha, S., Scherf, O., Bauer, W., & Bauer, S. (2008). Optimaler Fußgängerschutz durch situationsgerechte Einschätzung der Fußgängerbewegung. In *24. VDI / VW-Gemeinschaftstagung–integrierte Sicherheit und Fahrerassistenzsysteme (2008), of VDI-Berichte* (Vol. 2048, pp. 191–200). Düsseldorf: VDI Verlag.
14. Kolonko, M. (2009). *Stochastische Simulation* (1st ed.). Wiesbaden: Vieweg+Teubner.
15. Kroese, D. P., Taimre, T., & Botev, Z. I. (2011). *Handbook for Monte Carlo methods*. New York: John Wiley and Sons.
16. Niederreiter, H., & Spanier, J. (Eds.) (1998). *Proceedings of a Conference held at the Claremont Graduate University, Claremont, California, USA, June 22–26, 1998*. Berlin: Springer, ch. Monte Carlo and Quasi-Monte Carlo Methods.
17. Rubinstein, R., & Kroese, D. (2008). *Simulation and the monte carlo method* (2nd ed.). New York: John Wiley and Sons.

18. Shonkwiler, R., & Mendivil, F. (2009). *Explorations in Monte Carlo methods*. New York: Springer.

19. Helmer, T., Kühbeck, T., Gruber, C., & Kates, R. (2012). Development of an integrated test bed and virtual laboratory for safety performance prediction in active safety systems (F2012–F05-005). In *FISITA 2012 World Automotive Congress-Proceedings and Abstracts (2012)*. ISBN 978-7-5640-6987-2.

20. Helmer, T., Neubauer, M., Rauscher, S., Gruber, C., Kompass, K., & Kates, R. (2012). *11th International Symposium and Exhibition on Sophisticated Car Occupant Safety Systems*. Fraunhofer-Institut für Chemische Technologie ICT, 2012, ch. Requirements and methods to ensure a representative analysis of active safety systems (pp. 6.1–6.18). ISSN 0722–4087.

21. Kates, R., Jung, O., Helmer, T., Ebner, A., Gruber, C., & Kompass, K. (2010). Stochastic simulation of critical traffic situations for the evaluation of preventive pedestrian protection systems. In *Erprobung und Simulation in der Fahrzeugentwicklung*.

22. Benmimoun, M., & Benmimoun, A. (2010). Large-scale FOT for analyzing the impacts of advanced driver assistance systems. In *Proceedings of the 17th ITS World Congress*.

23. Dingus, T. A., Klauer, S. G., Neale, V. L., Petersen, A., Lee, S. E., & Sudweeks, J., et al. (2006). *The 100-car naturalistic driving study, phase II–results of the 100-car field experiment*. No. DOT HS 810 593. Washington: Department of Transportation, National Highway Traffic Safety Administration.

24. Gstalter, H. (1983). *Der Verkehrskonflikt als Kenngröße zur Beurteilung von Verkehrsabläufen und Verkehrsanlagen*. Dissertation, Technische Universität Carolo-Wilhelmina Braunschweig.

25. Reichart, G. (2002). Vom Fehler zum Unfall - ein neuer Ansatz in der Unfallforschung. In *VDA Technischer Kongreß 20002 - Sicherheit durch Elektronik (2002), VDA - Verband der Automobilindustrie* (pp. 59–76). Frankfurt: Main: VDA Verband der Automobilindustrie.

26. Bubb, H. (2001). Haptik im kraftfahrzeug. In T. Jürgensohn & K. P. Timpe (Eds.), *Kraftfahrzeugführung*. Berlin: Springer.

27. Green, M. (2000). How long does it take to stop? Methodological analysis of driver perception-brake times. *Transportation Human Factors, 2*(3), 195–216.

28. Snyder, M. B., & Knoblauch, R. L. (1971). *Pedestrian safety: The identification of precipitating factors and possible countermeasures*. Final Report DOT-HS-800-403. Washington: Departmant of Transportation, National Highway Traffic Safety Administration.

29. Consumer Information; New Car Assessment Program. (2008). No. NHTSA-2006-26555. National Highway Traffic Safety Administration.

30. Yasuki, T. (2005). *THUMS (Total human model for safety) in der Fußgängerschutz-Simulation*. Aachen: 14. Aachener Kolloquium Fahrzeug- und Motorentechnik.

31. Liers, H. (2010). *Extension of the Euro NCAP effectiveness study with a focus on MAIS3+ injured pedestrians*. Final report, VUFO GmbH.

32. Hannawald, L., & Kauer, F. (2004). *Equal effectiveness study on pedestrian protection*. Dresden: Technische Universität Dresden.

33. Ressle, A., Schramm, S., & Kölzow, T. (2010). Generierung von Verletzungsrisikofunktionen für Fußgängerkollisionen. In *Crash Tech 2010—Fahrzeugsicherheit 2020*.

34. Rosen, E., & Sander, U. (2009). Pedestrian fatality risk as a function of car impact speed. *Accident Analysis and Prevention, 41*, 536–542.

35. Zaloshnja, E., Miller, T., Romano, E., & Spicer, R. (2004). Crash costs by body part injured, fracture involvement, and threat-to-life severity, United States, 2000. *Accident Analysis and Prevention, 36*(4), 415–427.

36. Blincoe, L., Seay, A., Zaloshnja, E., Miller, T., Romano, E., Luchter, S., et al. (2000). *The economic impact of motor vechile crashes, 2000*. NHTSA Technical report DOT HS 809 446. Plans and Policy, National Highway Traffic safety Administration, May 2002.

37. Straube, M. (2010). *Volkswirtschaftliche Kosten durch Straßenverkehrsunfälle in Deutschland 2008*. Germany: Forschung kompakt 17/10, Bundesanstalt für Straßenwesen.

Chapter 4
Methodological Findings on Research on Driver Behavior

4.1 Objective

The driver and his behavior are of high relevance for the genesis of an accident (see Sect. 1.1, p. 1) as well as for the evaluation of changes in traffic safety by means of simulation (see Chap. 3). The change in driver behavior (e.g., particular reaction times) due to an active safety system can be derived by comparing use versus non-use of a system and serves as input for simulations as described in the previous chapter. To this end, driver behavior in response to a preventive pedestrian protection system (short: system) and his acceptance of specific system actions (especially false-positive responses) are investigated. The contents of this chapter are also partly included in [1, 2].

Due to the nature of preventive systems, which rely on information from environmental sensors, and the uncertainty of the situation itself (e.g., prediction of the pedestrian's movement before impact), a preventive protection system will produce false positives. The acceptance of false system actions by the driver is indirectly related to safety: If a system has a high rate of false positives (and therefore a very low acceptance by the driver), and if a possibility for deactivation of the system is available, the driver will possibly switch the system off and thus reduce the actual safety benefit to zero.

A key issue in the context of false-positive system actions concerns which parameters strongly influence the acceptance of the driver. These parameters can be considered especially during the design phases and assessed as far as possible in a simulation regarding frequency of occurrence. The "negative" aspect of false system actions as well as a low acceptance of those could be—at least partly—compensated for by the benefit of the system and the high acceptance of pedestrian protection itself. However, the *correct* action of such a system will hardly be experienced by any driver, as the probability of a pedestrian accident is extremely low (see Sect. 1.2). The relationship between level of hazard of the situation, as perceived by the driver, and plausibility of the system action is also an important issue.

© Springer International Publishing Switzerland 2015
T. Helmer, *Development of a Methodology for the Evaluation of Active Safety using the Example of Preventive Pedestrian Protection*, Springer Theses,
DOI 10.1007/978-3-319-12889-4_4

A subject study has been conducted in a driving simulator in order to investigate these key questions. The study focuses on acceptance of a preventive pedestrian protection system by the driver. A key methodological question is whether a highly critical accident situation including the subsequent accident can reliably and repeatably be conducted in a driving simulator. The key methodological question, whether a near-accident situation can reliably be reproduced in a driving simulator, is combined with a study on the potential safety benefit of a prototypical preventive pedestrian protection system.

The study was conducted in April and May 2010 using the dynamic driving simulator at BMW Group's Research and Innovation Center in Munich. Details regarding test design, subject sample, findings as well as further implications are discussed in this chapter.

4.2 Test Design and Subject Sample

4.2.1 Overall Design

The study was designed as a "naive" subject study using the dynamic driving simulator of the BMW Group in Munich. The subjects were neither informed about the specific research questions, i.e., the presence of a preventive pedestrian protection system, of the study nor the general setting. The official invitation announced a study on urban driving.

The preventive pedestrian protection system implemented in the experiment has the following key characteristics [2]:

- Optical pre-warning of the driver. Earliest at $TTC = 2.0$ s.
- Acoustical and optical warning. Earliest at $TTC = 1.5$ s.
- Automatic braking of the vehicle at 4 m/s^2. Earliest at $TTC = 0.9$ s.

The brake assist is reconfigured after the onset of a warning. In case of a driver-initiated braking after a warning, the desired deceleration is automatically set to 10.0 m/s^2 in order to achieve the maximum possible deceleration (the actual deceleration is dependent on the friction coefficient and the tire–surface combination). It is important to note in this context that a deceleration can only be realized with a time delay following the activation of the brake pedal or a request by the system. It can be assumed that a deceleration of, e.g., 10.0 m/s^2 needs about 0.3 s before the actual level of deceleration is reached [2].

An automatic action of the system only takes place if the driver has not reacted by evasive steering or braking before the corresponding TTC threshold. If the driver is already braking, but below the specified deceleration, the deceleration is increased by the system to the specified value. If the driver is already braking harder than the system would, driver-initiated braking will not be altered by the system.

The attention of the subjects should not be directed towards pedestrians during the test. To this end, all relevant situations were embedded into an interesting urban setting with frequent occurrence of pedestrians and considerable variety in the scene itself. The drivers had experienced substantial driving time without any hazardous (and especially noticeable) situations in order to bring them into a "normal" driving mode and to have a realistic chance to confront them "unprepared" with a traffic conflict. Thus, the presence of a pedestrian did not automatically imply any hazard or relevant situation. In this context, many non-critical (so-called "normal") situations were embedded in the test setting. Driving data as well as subjective ratings and comments were collected during the experiment.

The experiment lasted about 110 min and was structured into the following segments (the situations are described in detail below):

- Welcome; collection of demographical data.
- "Acclimatization" to the driving simulator ($\approx$12 min):
  - Start on a rural road, then transition to urban environment.
  - Two normal situations: Parking bay. Pedestrian crossing.
- Experimental section ($\approx$20 min):
  - Urban environment.
  - Four normal situations: Parking bay ($\times$2). Pedestrian crossing ($\times$2).
  - Two highly critical crossing situations with and without system. The sequence of those two situations is varied between subjects.
- Interview.
- Subjects are instructed regarding research questions and the specific functions of the system under investigation.
- Acceptance section ($\approx$22 min):
  - Different false system actions are presented to the driver.
  - The sequence is not varied between the subjects.
- Interview.
- "Clearance test": Artificial situation for investigation of the normal passing clearance for a pedestrian ($\approx$5 min).

The subject sample was reduced to 20 persons for the acceptance section and the clearance section. The subject sample itself is introduced after a discussion of the specific situations.

4.2.1.1 Test Design Specifics: Normal Situation

The normal situations cannot be identified by the subjects as experimental situations as they are implemented as common uncritical interactions with pedestrians without any system actions involved. The data collected are used to gain knowledge about the normal interaction and draw conclusions on the perception of discomfort or hazard while interacting with pedestrians.

In the *first normal situation*, the driver has to pass a pedestrian walking beside the street along a parking bay (Fig. 4.1). The lateral position of the pedestrian is different in each of the three occurrences of the situation in the experiment:

P-1. Walking on the road marking.
P-2. Walking left of the road marking (0.2 m left compared to P-1).
P-3. Walking right of the road marking (0.5 m right compared to P-1).

The *second normal situation* is a common crossing scenario where a pedestrian crosses the street from the right (see Fig. 4.2). A key parameter characterizing the mitigation of a potential conflict is the time-to-collision (TTC), which is defined here as the distance to the projected collision point divided by the current vehicle speed. The TTC when the pedestrian enters the street in this situation is varied during the experiment:

C-1. Low TTC: approx. 5.4 s.
C-2. Medium TTC: approx. 6.6 s.
C-3. High TTC: approx. 7.8 s.

The TTC values here represent the average for specific situations over all subjects. There is a small variation between the subjects due to a technical characteristic of the driving simulator: Pedestrians have a given (unalterable) motion characteristic where the starting point, the final speed, and the trajectory can be defined. That means that once the pedestrian is in motion, his speed cannot be adjusted depending on the current motion parameters of the vehicle. As a result, the actual TTC where the pedestrian steps on the street varies due to small deviations in vehicle speed between the subjects while approaching the pedestrian.

4.2.1.2 Test Design Specifics: Clearance Test

Twenty randomly chosen subjects participated in the clearance test. The task was to pass 10 pedestrians each. This task resembles an artificial (i.e., unrealistic) situation, where the pedestrians walk on an empty highway segment towards the vehicle (see Fig. 4.3). Each pedestrian has a different lateral clearance to the road marking. The drivers are instructed to drive at 50 kph and pass the pedestrian to the left at a clearance that is still acceptable for them and to ignore all road markings.

4.2.1.3 Test Design Specifics: Acceptance

The acceptance of false system actions was investigated in the second part of the experiment. All subjects were informed about the experiment and the system and were confronted with several situations that could trigger undesired system responses.

Fig. 4.1 Normal situation "passing" with variations P-1 to P-3 (from *top* to *bottom*)

Seven situations selected by an internal expert panel were presented to the subjects. The subjects had to give ratings

- regarding hazard of the undesired system action for the traffic situation as a whole and
- their individual acceptance of the false system action.

Fig. 4.2 Normal situation "crossing"

Fig. 4.3 Clearence test on the highway

The system response in those situations was triggered in order to obtain a reliable presentation for every subject. In order to get a more realistic feeling in some situations, the TTC of the optical pre-warning was set earlier than described above. After each situation, the subjects had to stop the vehicle and had to go through an interview. The situations are described in detail in the following:

Situation 1 (Fig. 4.4): False system response due to a pedestrian on a *traffic island*: The driver approaches an urban environment at approx. 60 kph. The system reacts because of a pedestrian standing at the edge of a traffic island. The pedestrian is directly in front of the vehicle at the moment of the warning.

Fig. 4.4 Acceptance
situations: traffic island (1);
evasive maneuver (2); curve
(3), (from *top* to *bottom*)

Situation 2 (Fig. 4.4): False system response while negotiating an *evasive maneuver* due to a pedestrian on the left side of the street: The driver has to make an evasive maneuver because of a parked car in his lane. The system reacts because of a pedestrian standing at the edge of the left sidewalk.

Situation 3 (Fig. 4.4): False system response due to a pedestrian on the left side of the street while negotiating a right *curve*: The driver follows the street in a curve to the right. The system reacts because of a pedestrian walking on the left sidewalk.

Fig. 4.5 Acceptance
situations: intersection curve
(4); T-intersection (5);
parking bay (6), (from *top* to
bottom)

Situation 4 (Fig. 4.5): False system response due to a pedestrian on the opposite
side of an *intersection* while negotiating a right *curve*: The driver follows the street
in a right curve while crossing an intersection.

Situation 5 (Fig. 4.5): False system response due to a pedestrian on the opposite
side of a *T-intersection*: The driver approaches a T-intersection and the pedestrian is
standing on the opposite sidewalk directly in front of the vehicle.

Fig. 4.6 Acceptance situation: pedestrian running at intersection

Situation 6 (Fig. 4.5): False system response due to a pedestrian walking at the side of the street in a *parking bay*: The driver passes the pedestrian and the system reacts.

The situations described above represent undesired actions of the system. The next situation does not present an undesired system action. In this situation, the driver makes a right turn at an intersection while a pedestrian is running across the street from behind (Fig. 4.6). The system does not show any reaction; this is a situation with a false-negative response. The question here is whether the subjects expect the system to handle this situation or not.

4.2.1.4 Test Design Specifics: Highly Critical Situation

The last situation included in the experiment was a highly critical situation. The subject drives for a long time through the city and has to fulfill a secondary task several times without anything happening. At the end of the first half of the experiment, the driver is confronted with a highly critical situation, which represents the most common real-world accident scenario (see [3] or Sect. 3.2, p. 52). A pedestrian is crossing the street from the right (Fig. 4.7). The presence of other pedestrians could have an influence on the situation, but is not recorded in accident data. At that time, the driver is working on the secondary task (see description below). The situation has to be managed twice by each driver (with and without system). The sequence of the highly critical situation was varied between the subjects (regarding the system), i.e., 20 persons had the first highly critical situation with the system, the other 20 without. The actual number of measurements per situation can vary as not every person went through the whole experiment.

Fig. 4.7 Highly critical situation: pedestrian crossing from the right

4.2.1.5 Test Design Specifics: Secondary Task

The secondary task is a visual loading task, which does not force the driver to take his hands off the steering wheel. The loading task has been successfully used in previous internal studies. It is not interruptible and produces a constant visual distraction. The loading is constant and it is hardly possible to develop individual strategies for solving the task. The driver is confronted with single letters displayed in the central information display of the vehicle. Each letter is displayed for a very short time, which makes constant monitoring by the driver necessary. As soon as a number appears instead of a letter, the driver has to press a button on the steering wheel within 1 s. The beginning of the task is introduced by an acoustical signal. The task appears several times within the experiment and lasts about 1min. each time. The task is used to produce visual distraction and "prepare" the subject in a better way for the critical situation.

4.2.1.6 Subject Sample Characteristics

The subject sample consisted of 40 persons, aged 22–60 years (average 37.3 years, SD 10.9 years). 13 persons were female, 27 male. All subjects were BMW Group employees not working or acquainted with driver assistance or systems of active safety.

4.3 Acceptance of the System in Specific Situations

The acceptance of a preventive system is an important criterion during development. As long as the future driver of a vehicle has the choice to decide whether to have or not to have a system, his acceptance will influence his decision to use or deactivate the system (if possible) or buying it again. As a consequence, acceptance does influence the safety benefit of a system, especially, if the system can be deactivated (in this case due to low acceptance). This section explains the findings on acceptance from the driving simulator study.

Acceptance is not only limited to undesired system actions, e.g., false positives or false negatives, but also includes desired actions (e.g., correct positives). Two different approaches are used to gather information about the drivers' acceptance:

- The first is whether to determine the level of discomfort or perceived hazard while interacting with pedestrians without any system. This level can be used as a threshold, separating a system action which is desired or understood as correct and helpful by the driver from one which is regarded as unnecessary or annoying. Key characteristics are the TTC at the beginning of an (uncritical) evasive maneuver, the TTC at the beginning of an (uncritical) braking maneuver, and the lateral passing clearance.
- The second approach focuses on typical situations involving different false-positive system actions as well as a false-negative system action (i.e., missed alarm). These are the acceptance situations mentioned in the previous section. Key characteristics are subjective ratings on acceptance of the false alarm and perceived hazard of the situation.

The results include lateral and longitudinal behavior. The lateral component is evaluated using the normal situation "passing" and the clearance test. The drivers were instructed to drive 50 kph and pass the pedestrian. The lateral passing clearance is defined as the distance between the side of the vehicle and the near side of the pedestrian (assumptions: vehicle width is 1.79 m and pedestrian width is 0.60 m). Figure 4.8 shows that the pedestrian is passed in an average lateral clearance between 1.43 and 1.60 m (SD 0.23–0.35 m). The smallest variation, referring to the standard deviation, is seen in the situation "walking right of the road marking" and the largest in the situation "walking left of the road marking"; nearly as large as in the clearance test. The last effect can be attributed to the synthetic nature of the clearance test. The differences in the mean are non-significant, using t-tests (situations: 1 versus 2: $t = 1.26$; 2 versus 3: $t = 0.94$; 1 versus 3: $t = -0.03$). For an introduction on t-tests see Sect. 5.2.5, p. 97.

The drivers also made an evasive maneuver to achieve the lateral clearance as well as changed their longitudinal behavior. The pedestrian was visible from far away, so the situation was regarded as uncritical and the drivers had as much time to react and adapt to the situation as they liked. Figure 4.9 gives the TTC at the beginning of a steering reaction. The persons started to steer on the average at a TTC between 2.7 and 3.0 s. In condition P-2 (i.e., the lowest lateral clearance of the pedestrian to

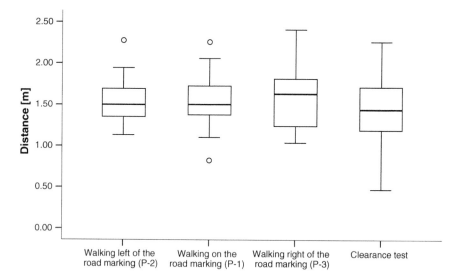

Fig. 4.8 Lateral clearance between vehicle and pedestrian in the normal situations "passing" and the clearance test. The *circles* in the boxplots indicate near outliers

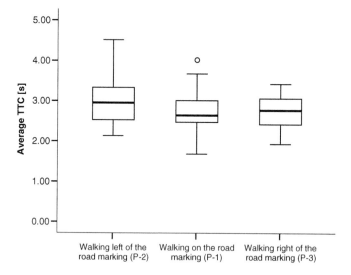

Fig. 4.9 TTC to the pedestrian at the beginning of the steering action in the normal situations "passing"

the street) the drivers reacted 0.3 s earlier (SD 0.42–0.62 s). The differences in the mean are non-significant, using t-tests (situations: 1 versus 2: $t = -1.87$; 2 versus 3: $t = -0.56$; 1 versus 3: $t = -1.48$).

The longitudinal reaction in the "passing" situations depended highly on the position of the pedestrian. A braking reaction is observed in

- 1 of 29 situations for condition P-3 (i.e., largest initial clearance);
- 7 of 35 situations for condition P-1 (i.e., middle initial clearance);
- 19 of 34 situations for condition P-2 (i.e., smallest initial clearance).

In order to have sufficient data for interpretation only condition P-2 was used for analysis of longitudinal driver reactions. The drivers started braking on the average at a TTC of 4.10 s (SD = 0.61 s).

The longitudinal behavior was further investigated using normal situation "crossing". The brake response in this situations is also dependent on the initial TTC of the situation itself.

- At the lowest TTC (approx. 5.4 s; condition C-1), 35 of 35 persons braked.
- At the medium TTC (approx. 6.6 s; condition C-2), 18 of 38 persons braked.
- At the largest TTC (approx. 7.8 s; condition C-3), 12 of 35 persons braked.

The initial reaction, i.e., taking the foot away from the accelerator pedal, was hard to evaluate, as many persons in this simulator experiment did not constantly apply the accelerator pedal. As a consequence, the results could be distorted by simulator artifacts and are therefore not discussed. The start of braking is a valid indicator for this situation and is shown in Fig. 4.10. The average TTC is 4.07 s (SD = 0.90 s) for condition C-1 and 3.82 s (SD = 1.65 s) for condition C-2. Due a low number of measurements (i.e., braking reactions), the results presented exclude condition C-3.

The results shown above allow for an interpretation concerning acceptance of possible system actions. The findings are consistent and show a stable lateral passing clearance around 1.5 m independent of the initial conditions or the test setting. The steering reaction is consistent and shows a reaction around 3 s TTC. Start of braking as

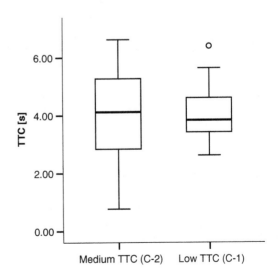

Fig. 4.10 TTC to the pedestrian at the beginning of the braking reaction in the normal situations "crossing"

longitudinal reaction was observed around 4 s TTC. As the situations were uncritical and the drivers had as much time to react as they liked (the pedestrians were visible long before the TTC values mentioned above), it can be concluded that these values indicate a comfort zone which the drivers like to maintain. It does not mean that they feel uncomfortable immediately below those values or that the situation is regarded as hazardous immediately below those values.

It can be concluded from the findings in the normal situations that a system configuration as described above will be accepted quite well. The system reaction (i.e., acoustical warning at a TTC of 1.5 s or automatic braking at a TTC of 0.9 s) is at a TTC where nearly all drivers would have reacted (using the results presented above) if they had the chance to perceive the pedestrian. This is also confirmed by the results of the interview regarding the highly critical situation (see next section). It can also be concluded that at higher TTC levels than discussed using Fig. 4.10, nearly no drivers feel a necessity to react to the situation.

The next part focuses on the investigation of the acceptance situations mentioned above. The research question is the subjective perception of hazard and acceptance of false system actions in these situations. The subjects were suddenly confronted with the situations. Out of the 20 subjects, 10 got false warnings and an automatic braking (in case the TTC values became small enough) and the other 10 got only a warning without an automatic braking as system response. The system response was triggered in the situations to get a reliable presentation for as many subjects as possible. This part of the experiment has the characteristic of a presentation and is meant to produce qualitative insights into acceptance of false system actions, not to produce an amount of data that is statistically usable. A system ready to go into mass production would handle most of these situations by, e.g., predicting the trajectory of the vehicle, predicting the pedestrian's movement or calculating a collision probability.

The first rating uses a 100 % scale to investigate the perceived *hazard of the situation* (where 100 means maximum hazard and 0 minimum hazard). The subjects were instructed to rate the hazard for the *whole* surrounding traffic, not for the pedestrian alone. The rating regarding hazard of the situation is displayed in terms of the median in Fig. 4.11, and the corresponding ranks are given in Table 4.1 for both conditions (i.e., warning only and full system response). Whereas the absolute value in this scale cannot be interpreted, the relative differences do have a meaning. The situations "curve", "traffic island", and "intersection curve" were rated as most hazardous (followed closely by "parking bay", "evasive maneuver", and "T-intersection"). The subjects explained this by the unpredictability of the system reaction ("curve" and "intersection curve") and by the current speed of the vehicle ("traffic island"). Comparing both conditions, the rating differs most for "curve" and "intersection curve". A possible explanation is that automatic braking irritates the subjects more while negotiating a curve than a straight road. It can be seen in every situation that automatic braking is regarded as more hazardous than only a warning (exception: "T-intersection").

The second rating focuses on *acceptance* of the undesired system response. It also uses a 100 % scale, where 100 stands for lowest acceptance and 0 for highest acceptance. As explained before, only relative differences do have a meaning. Figure 4.12

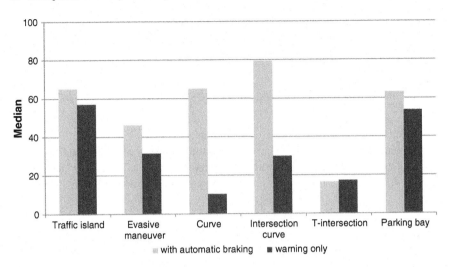

Fig. 4.11 Rating "hazard of the situation". Medians

Table 4.1 Rating "hazard of the situation": Ranks

	Traffic island	Evasive maneuver	Curve Curve	Intersection curve	T-intersection	Parking bay
With automatic braking	2[a]	4	2	1	5	3
Warning only	1	3	6	4	5	2

[a] "Traffic island" shares the same rank with "Curve"

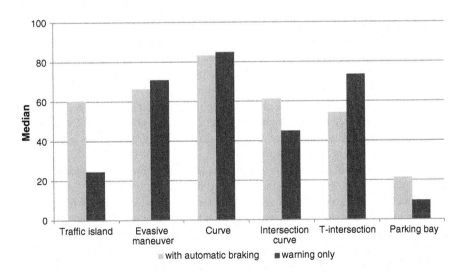

Fig. 4.12 Rating "acceptance of false system action". Medians

Table 4.2 Rating "acceptance of false system action": Ranks

	Traffic island	Evasive maneuver	Curve Curve	Intersection curve	T-intersection	bay bay
With automatic braking	4	2	1	3	5	6
Warning only	5	3	1	4	2	6

gives the medians and Table 4.2 the ranks for the situations and both conditions. The lowest acceptance can be observed in the situations "curve", "evasive maneuver", and "T-intersection". The subjects provided two explanations for this:

- The pedestrian does not move and thus a real hazard, i.e., the need for a system action, cannot be seen.
- The pedestrian is on the far side of the street, and the resulting hazard for the pedestrian is also seen as low.

There are large differences between the conditions *full system* and *warning only*, especially in the situations "traffic island" and "parking bay". Braking in front of the traffic island surprises and confuses the subjects. Automatic braking while passing the pedestrian next to the parking bay is also highly unacceptable compared to warning only, as the pedestrian has nearly been passed by the time the braking sets in.

The last acceptance situation (i.e., running pedestrian at intersection) has to be interpreted separately, as it does not present a false system action but a miss (i.e., false negative). The research question was whether the subjects expected the system to act in this situation or not. 12 out of 20 subjects did not expect the system to act in this situation. The rest mentioned that adequate sensors would be able to detect the pedestrian, and thus a system response would be possible. The rating revealed that the hazard of the situation is regarded as low, and acceptance of the lack of an action is quite high. This can be explained by a relatively low vehicle speed while turning and a high attentiveness to the surrounding traffic by the driver.

The acceptance part of the experiment can be summarized as follows. The normal situations produced stable and consistent results giving the lateral clearance while passing a pedestrian, the TTC values while approaching a pedestrian or conducting an evasive maneuver. The results give a strong indication about the upper boundary for system actions regarding acceptance. If a system acts at higher TTC thresholds or a greater lateral clearance, the drivers will most likely have a low acceptance, because they would not see a need to react.

The presentation of false system action gives information about influencing factors for the perceived hazard of a traffic situation and the acceptance if a false system action occurs. For the perceived hazard, these factors seem to be predictability of the false system action and vehicle speed. For the acceptance of false system actions, the movement and position of the pedestrian relative to the vehicle are important. Acceptance also decreases with increasing vehicle speed.

As the questions investigated can have an influence on the safety benefit of the system, the intervention strategy of the system need to be designed accordingly.

4.4 Driver Behavior in Highly Critical Situations

The first part of the experiment included a highly critical situation. The research question is whether a near-accident situation can be reliably reproduced in the driving simulator using a test design as described. In case the situation works for most of the subjects, a change in driving behavior due to the preventive pedestrian protection system (as described above) can be quantified. A between-subjects design is used for the experiment. Half of the subjects experience the situation with the preventive system first, the other half without the system. The next time the situation is repeated with changed experimental conditions. It is important to note that the results from the second situation could be distorted by the fact that the subjects are not unprepared anymore. After the first highly critical situation, their behavior could change. As a consequence, the results are given for the first situation.

As the experimental conditions were varied between subjects, the first step is a comparison of the initial conditions while entering the situation. The subjects were instructed to drive 50 kph. The actual initial speed (independent of the experimental situation) was 51.92 kph (SD 4.10 kph) without system versus 52.39 kph (SD 4.80 kph) with system. The differences in the mean are non-significant using the T statistic ($t = -0.41$); the initial vehicle speeds are thus comparable. The TTC when the pedestrian reaches the curb was on the average 1.90 s (SD 0.26 s) (only first situation). As discussed in the previous section, the pedestrian cannot be controlled dependent on the movement of the car. As a consequence, the TTC at the curb shows some variation between the subjects.

The following reactions in this situation are possible in general:

- The driver perceives the pedestrian and reacts before the system does. In this case, the system is deactivated.
- The driver gets the optical pre-warning and reacts.
- The driver gets the pre-warning and the acoustical warning and reacts.
- The last possible reaction of the system, an automatic braking, was not observed in the experiment.

Key findings include the time of driver reaction with respect to the system action, the TTC at the activation of the brake pedal, the time between the acoustical warning and the activation of the brake pedal, and the maximum deceleration. In addition the perception of the different warning signals by the drivers was investigated.

Figure 4.13 shows the distribution of driver reactions in the first situation. 8 of 19 drivers braked after the two warnings. How critical the situation became during the test is indicated by the number of actual collisions. Considering the first situation only,

- 2 of 19 drivers had a collision using the system and
- 2 of 18 drivers had a collision not using the system.

These numbers indicate that there is no difference in the result due to the experimental conditions. The impact speed cannot be evaluated due to low number of collisions. The distance to the pedestrians in all avoided collisions is comparable for both

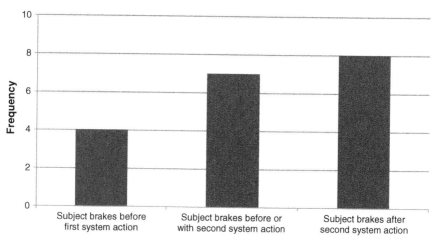

Fig. 4.13 Distribution of driver reactions in the first situation

conditions (non-significant difference in the mean, $t = -0.20$). As these results seem surprising given the design of the experiment and functions of the system, possible reasons are explained and discussed in the following.

The situation was constructed based on the relevance of real-world accidents (see Sects. 3.2 and 4.2). The subjects were visually distracted by a secondary loading task, which was calibrated using expert knowledge from previous experiments. The whole timing of the situation was optimized in a pre-test with several experts and dummy subjects.

Figure 4.13 shows that 11 of 19 subjects reacted before the acoustical warning. Figure 4.14 gives the TTC to the pedestrian at the activation of the brake pedal. Regarding the first occurrence of the situation, the driver reacted on average a bit earlier with the system (1.55 s versus 1.44 s), but this difference in the mean is non-significant ($t = -0.88$). Figure 4.15 gives the duration in time from the onset of the acoustical warning (at an earliest TTC of 1.50 s) and the activation of the brake pedal. As the maximum is 0.075 s, it is obvious that the driver reaction is *no* reaction to the acoustical warning. It must be concluded that the driver observed the pedestrian earlier during the test and decided to brake. Whether the reaction was triggered by the optical warning cannot be assessed with certainty. Although theoretically possible (as the optical warning is given at an earliest TTC of 2.0 s and so takes place 0.5 s before the acoustical one) it seems unlikely that this explains the moment of reaction of the driver, since only five subjects reported that they had observed the optical warning.

Considering the reaction of the drivers itself, the following can be stated. All drivers applied the brakes in reaction to the imminent danger. Some drivers also made minor changes in their lateral position. The brake reaction itself is of interest, as the system includes a brake assist, which gives the driver an acceleration of $-10.0 \, \text{m/s}^2$ in case a warning was issued and the brake pedal was activated. Figure 4.16 shows

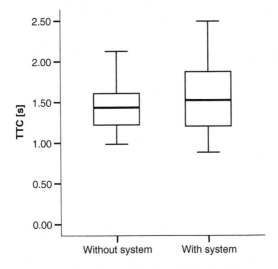

Fig. 4.14 TTC to the pedestrian at activation of brake pedal by the drivers

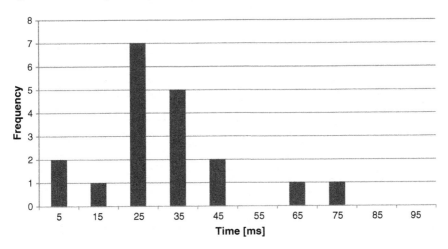

Fig. 4.15 Time between acoustical warning and activation of the brake pedal by the drivers

the maximum deceleration for the first situation for both experimental conditions. Regardless of the condition, nearly all drivers were able to realize a deceleration beyond $9 \, m/s^2$; without system the mean was $10.39 \, m/s^2$ (SD $0.60 \, m/s^2$) and with system $10.07 \, m/s^2$ (SD $0.63 \, m/s^2$).

Accident statistics reveal that drivers do not realize the maximum possible deceleration even in accidents ($3.85 \, m/s^2$, SD $3.33 \, m/s^2$). The accident statistics thus indicate that a brake assist, as implemented in this system, has a great potential (Fig. 4.17).

A second effect is connected to this phenomenon. The duration from beginning of brake pedal activation to the maximum deceleration is also non-significantly different

Fig. 4.16 Maximum deceleration in the first situation, stratified by experimental conditions. *Circles* indicate near outliers and *stars* far outliers

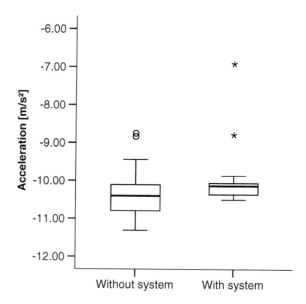

Fig. 4.17 Mean deceleration in pedestrian accidents [4]

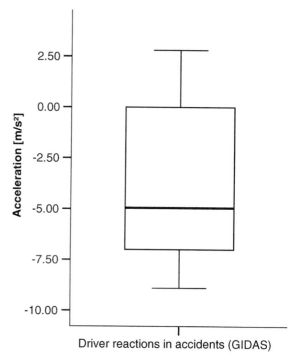

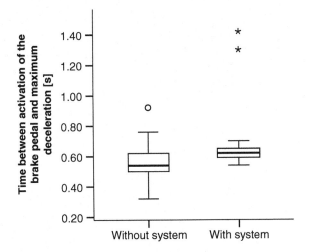

Fig. 4.18 Time from beginning of brake pedal activation to maximum deceleration

considering the means ($t = -2.01$), see Fig. 4.18. The experiment does not reveal the expected effect of the preventive system regarding maximum deceleration and time for building up the maximum deceleration. A difference between driving simulator and real vehicle explains these effects. Drivers tend to avoid the hard braking maneuvers in reality even in accident situations, as they try to avoid high decelerations. Due to technical limitations, the driving simulator scales the decelerations experienced by the subjects (and all other accelerations as well) to a lower level, whereas the kinematic deceleration, as implemented in the vehicle dynamics model of the simulator, is realistic. As a consequence, drivers tend to realize far higher decelerations in the simulator than they would in reality. Due to these circumstances, the results regarding deceleration obtained in this experiment cannot be used directly for the evaluation of the effects of the preventive pedestrian protection system or as input for a driver model as needed in simulations.

The basic idea behind this situation was to present a realistic pedestrian accident scenario (crossing from the right, no visibility obstruction, urban setting, daylight, …) in the driving simulator. The drivers went through a long period of driving without any special event, were not informed about the research questions, and were visually distracted by a secondary task during the highly critical situation. The objective for this situation was to test whether an accident situation can be reproduced in a stable way in such an experimental environment. The difference to other studies is that not a critical situation but an accident situation should be created. As the results show, it was not possible to bring the subjects reliably and repeatably into the critical accident situation (TTC at braking should have been at least below 1.0 s TTC) as only a few accidents did happen under baseline conditions.

Several possible explanations and ideas for further studies have been developed. The subjects probably found a strategy to work on the secondary task *and* react properly to the traffic situations. A possible design change would be to further increase the level of distraction. The experimental situation itself, i.e., taking part in an experiment in the driving simulator, could additionally have influenced the subjects as well as the perception of the environment and the traffic situation in the simulator. Another technical possibility would be a visibility obstruction of the pedestrian. This deviates from the realistic scenario as described in Sect. 3.2, but could help bring the subject into the accident situation and thus be an experimental necessity.

4.5 Conclusion

The experiment in the dynamic driving simulator as described above gives insight into driver behavior regarding pedestrians, the issue of acceptance of a preventive pedestrian protection system, especially during the presentation of false system actions, and the methodological challenges when investigating realistic accident scenarios in a driving simulator. As the age characteristics of the sample used here range from 22 to 60, an interesting extension of this study might especially consider a population of older drivers, due to their possibly changed perceptual and reaction characteristics.

Everyday normal situations were used to assess the levels of discomfort or hazard while passing a pedestrian walking in the same direction as the vehicle moves or approaching a crossing pedestrian. The drivers could perceive the pedestrians from a great distance, passed them at an average lateral clearance of 1.5 m and started to brake on average at a TTC of 4 s. The implication of these findings is that a system issuing warnings within these boundaries (laterally and longitudinally) has a high chance of acceptance, if the driver did not react himself in advance. The consequence of low acceptance could be a deactivation of the system (if possible) or reluctance to purchase the system in a future vehicle (if optional equipment). In both cases the safety benefit would be negatively affected by low acceptance.

The acceptance ratings revealed that the subjects regarded situations as especially *unacceptable* where an endangerment of the pedestrian was not obvious to them. The reasons given by the subjects were that the pedestrian is not moving or is far away from the current position of the vehicle or its present trajectory. Situations were regarded as *hazardous* for the surrounding traffic when false system actions were unpredictable for the drivers, the situations included higher vehicle speeds or the situations involved complex maneuvers. If the drivers' attention was already high in a situation, a false system action was regarded as less hazardous.

An important finding is that the investigation of highly critical situations in the driving simulator proves to be challenging. The drivers reacted about 0.1 s earlier (difference is non-significant) at an average TTC of 1.44 s with the preventive pedestrian protection system installed than without. All drivers observed the pedestrian and reacted before or nearly at the time of the acoustical warning. The braking reaction in terms of timing, maximum deceleration, or duration between initiation of the braking and the maximum deceleration could not be evaluated regarding system effectiveness, as the driving simulator does not allow for an interpretation of these measurements. The missing realistic kinesthetic feedback due to the technically necessary scaling of the real accelerations is responsible for the magnitude of the braking reactions, not the experimental conditions. The drivers brake much harder than in real accidents and also tend to push the brake pedal very fast, which is also suspected by experts not to be the case in real critical situations.

Overall, the highly critical situation hardly led to any accidents in the baseline condition, which gives an indication about the challenges of bringing the subjects into the situation in a driving simulator. These results shows that the realistic and reliable construction of an *accident* situation (in contrast to *critical* situations as used in many other experiments) is challenging even if a tested secondary task and an optimized test design are used. Possible solutions are a stronger distraction of the subjects or a visibility obstruction of the pedestrian. Clearly, more research regarding the methodology of subject testing in accident situations is needed.

The technological limitations of the driving simulator as a method in general as well as of the specific one used here become evident when evaluating brake reactions. On the one hand, the time series of braking itself cannot be evaluated. On the other hand, the driving simulator is the method of choice to do subject testing in accident or critical situations.

Since this experiment had the driver in the focus, it is obvious that the behavior of the pedestrian is also an important field of research, although not part of this thesis. His actions do influence the situation itself as well as the system actions (e.g., prediction of collision probability).

A combination of findings obtained in different kinds of experiments—for example, using processes and techniques as described in Sect. 3.1—is necessary to get a complete picture of the effects of a preventive system involving the vehicle as well as the driver.

References

1. Helmer, T., Ebner, A., Jung, O., Paradies, S., Huesmann, A., & Praxenthaler, M. Fahrerverhalten in Fußgängersituationen mit und ohne Unterstützung eines präventiven Sicherheitssystems - Herausforderungen bei der empirischen Bewertung. In *AAET 20011 - Automatisierungssysteme, Assistenzsysteme und eingebettete Systeme für Transportmittel*. Gesamtzentrum für Verkehr Braunschweig e.V., 2011.
2. Praxenthaler, M. (2011). Nutzertest Präventiver Fußgängerschutz. Abschlussbericht, empirience
3. Ebner, A., Samaha, R. R., Scullion, P., & Helmer, T. (2010). Methodology for the development and evaluation of active safety systems using reference scenarios: Application to preventive pedestrian safety. In *Proceedings of the International Research Council On Biomechanics of Injury (IRCOBI)* (pp. 155–168), (2010).
4. German In-Depth Accident Study: Unfalldatenbank 07.1999–12.2008. Dresden, Hannover, 31.12.2008.

Chapter 5
Probabilistic Modeling of Pedestrian Injury Severity

5.1 Objective and Research Questions

This chapter presents the methodology necessary for the construction of evidence-based probability models for pedestrian injury severity in frontal vehicle crashes using empirical, in-depth accident data. The primary aim is thus to apply statistical methodology in order to estimate models predicting injury severity and mortality of pedestrians involved in vehicle crashes, based on the conditions of impact. The results are intended to improve and quantify the predictability of pedestrian injury severity during design and development phases of preventive pedestrian protection systems as well as to provide a basis for comparison with safety benefits due to measures of passive safety. The data sets used are the German In-Depth Accident Study (GIDAS) [1] and Pedestrian Crash Data Study (PCDS) [2] for the US.

It is well established that collision speed is the most important predictor for injury severity [3–7]. However, for constructing probability models for advanced applications, several additional research questions arise:

1. Which injury scale available in the data sets would be most suited for deriving probabilistic models?
2. Do multivariate models provide a better prediction than univariate models based solely on impact speed?
3. Does a splitting into subgroups defined by pedestrian age provide a better prediction than models comprising all ages?

The *first* of these issues refers to the choice of a metric to describe the injury scale. In the data sets considered here, pedestrian injuries were originally coded according to the Abbreviated Injury Scale (AIS), revision 90 [8–10] (for cases 2008 and newer GIDAS also includes AIS coding following the 2005 revision [11–13]). Table 5.1 gives the AIS levels as well as the lethality rate associated with each level. The maximum AIS value (MAIS) of a person is separately coded and serves as an indicator for overall injury severity.

© Springer International Publishing Switzerland 2015   91
T. Helmer, *Development of a Methodology for the Evaluation of Active Safety using the Example of Preventive Pedestrian Protection*, Springer Theses,
DOI 10.1007/978-3-319-12889-4_5

Table 5.1 AIS codes and description [2] with corresponding lethality rate [14]

AIS	Severity description	Lethality rate [%]
0	Not injured	0.00
1	Minor injury	0.00
2	Moderate injury	0.07
3	Serious injury	2.91
4	Severe injury	6.88
5	Critical injury	32.32
6	Maximum (untreatable) injury	100.00

Another established injury coding scale is known as the Injury Severity Score (ISS) [9, 10, 15, 16]. The ISS is defined as the sum of the squares of the highest AIS scores in each of the three most severely injured body regions (out of six regions in total). It ranges from 0 to 75; 75 is the maximum and is defined if at least one body region has an AIS of 6. There are strong indications in the medical literature that ISS gives a more precise estimate of the overall injury severity than MAIS [17, 18].

The *second* research question refers to the number of variables included in the models. As stated above, impact speed is the most important predictor for injury severity and mortality. However, it is known, for example, that fatality risk can be predicted more precisely using pedestrian age in addition to impact speed [6]. Thus, considering the spectrum of variables coded in the databases, it is important to identify potential explanatory variables beyond impact speed that could improve the predictive accuracy of the models. Possible explanatory variables include vehicle kinematics (e.g., collision speed), vehicle characteristics (e.g., height of the front bumper), and pedestrian physiology (e.g., age).

The *third* research question takes the biomechanical differences due to pedestrian age into account. It is well known that the biomechanical response with respect to injury severity is dependent on age [19–21]. To this end, this study will also investigate whether a splitting of the population into subgroups depending on pedestrian age improves the quality of injury modeling.

In addition, two constraints concerning injury probability models are implemented. The first one is a simple definition: the injury or fatality probability for an impact speed $v_c = 0$ kph is defined as zero. The second one is more subtle. In the case of several cumulative outcome categories (e.g., ISS9+, ISS16+, and ISS25+), the probability for a larger outcome category, e.g., ISS16+, must be at least as large as the probability for another smaller set, e.g., ISS25+ ($p_{ISS16+} \geq p_{ISS25+}$), which itself is a subset of the first one ($ISS25+ \subseteq ISS16+$). If that constraint has not been taken into account explicitly in model development, then it needs to be tested to guarantee plausibility. To this end, a conditional probability simulation is introduced which generates synthetic vehicle-to-pedestrian accidents with all input parameters necessary for the models in question. These two constraints are investigated, tested, and their implications are discussed together with remarks for correct implementation of the models. A new methodology of constructing probability models for several cumulative outcome categories (e.g., ISS0-8, ISS9-15, ISS16-24, and ISS25+) by means of conditional probabilities is developed and tested for the constraints.

5.2 Data and Statistical Methods

5.2.1 Study Data Characteristics

The focus is on two data sets in the following analysis: the German In-Depth Accident Study (GIDAS) and the Pedestrian Crash Data Study (PCDS) from the US.

The German In-Depth Accident Study contains a sample of accidents collected from 1999 by the Hannover Medical School (MHH)and the Technical University of Dresden (TUD). The project is coordinated by the Federal Highway Research Institute (BASt) and the Research Association of Automotive Technology (Forschungsvereinigung Automobiltechnik, FAT). The companies represented by the FAT for this project are: Ford-Werke GmbH, Volkswagen AG, Daimler AG, BMW AG, General Motors, Dr. Ing. h.c. F. Porsche AG, Autoliv Inc., TRW Automotive, and Johnson Controls Inc. The data are collected in two geographical areas including the cities of Hannover and Dresden in Germany. The sampling includes only cases with personal injury. The accident investigation follows a shift plan. Collected cases are compared and weighted to the federal statistics every year. The collected cases are considered representative for the sampling area and can be regarded as nationally representative, if regional influences can be neglected for the specific research question (which is the case for most aspects of passive safety). This procedure leads to the collection of about 2,000 cases annually, including acquisition of information regarding the site of an accident, inspection of the vehicle, detailed medical information and extensive reconstruction [22].

Ten years of data from GIDAS have been used for the present analysis (07/1999 to 06/2009; Version Dec. 2009) resulting in 16,827 fully reconstructed cases. The following selection and exclusion criteria have been applied to the data set for this study:

- Accidents with pedestrians involved: $n = 2,270$.
- Primary collisions with passenger vehicles at vehicle front (including roof): $n = 1,073$.
- Collisions with pedestrians of age four and older, since the actual position of infants at the moment of impact is not coded: $n = 998$.
- Impact speed of the vehicle available, as it is by far the most important predictor [4–6]: $n = 915$.

The basic data set contains 915 cases available for analysis.

The US Pedestrian Crash Data Study (PCDS) contains a sample of pedestrian accidents between 1994 and 1998, generated by the Transportation Data Center at the University of Michigan's Transportation Research Institute (UMTRI) for the National Center for Statistics and Analysis (NCSA) of the National Highway Traffic Safety Administration (NHTSA). A total of 552 cases were collected in the following cities: Chicago, Illinois; Buffalo, New York; Fort Lauderdale, Florida; Dallas, Texas; Seattle, Washington, and San Antonio, Texas. Most of the cases are not included in the National Automotive Sampling System (NASS), Crashworthiness Data System (CDS) or the NASS General Estimates System (GES). The main selection and exclusion criteria for PCDS applied during data collection may be summarized as follows:

- Pedestrian accidents only (no cyclists, etc).
- Impact between one vehicle and one pedestrian.
- Cars and light trucks only.
- Vehicles of model year 1990 or later.
- Initial contact of the vehicle with the pedestrian was in front of the A-pillar.
- Vehicle part striking pedestrian was (undamaged) original equipment.

Due to the sampling scheme and selection criteria, the PCDS data set might not constitute a representative sample of pedestrian accidents for the US as a whole regarding all characteristics. However, the PCDS does seem to be quite representative with regard to the frequency distribution of accident scenarios [23, 24]. In any case, it is quite useful for the intended purpose of identifying risk factors and estimating predictive risk models in the accident classes considered [2, 25, 26].

Data were filtered for frontal vehicle impacts with a pedestrian of age four and older, resulting in 450 collisions. In the following analysis, only cases with impact speed available were considered (see above), resulting in 369 collisions.

5.2.2 Coding of Target Variables

In both databases, pedestrian injuries in each accident were originally reported and scored according to AIS for each separate injury. Based on these AIS scores, MAIS and ISS, which are both ordinal measures of severity, were then calculated and coded. Meaningful levels for MAIS as binary target variables suited for logistic regression are MAIS2+ (MAIS $\geq$ 2, at least moderate injury), MAIS3+ (MAIS $\geq$ 3, at least serious injury), MAIS4+ (MAIS $\geq$ 4, at least severe injury), MAIS5+ (MAIS $\geq$ 5, at least critical injury).

Following Hakkert et al. [27] in the target variable definition, ISS is coded as binary target variables ISS9+ (ISS $\geq$ 9, at least moderate severity), ISS16+ (ISS $\geq$ 16, at least serious injury), ISS25+ (ISS $\geq$ 25, at least very serious injury). The consistency of the data regarding MAIS and ISS was checked by computing the ratio of squared MAIS to ISS, which must be within 0.33 (i.e., three body regions have an injury whose severity equals the MAIS of that person) and 1.0 (i.e., only one injury per person). In addition to the medical literature mentioned above, the use of the ISS is supported by the fact that 90.8 % of the pedestrians in frontal impacts in PCDS have at least two injuries (mean = 8.7; standard deviation = 8.6).

Fatalities were coded and investigated independently, as they are distributed over a range of ISS or MAIS values [6]. Cases with missing injury or mortality data were excluded from analyses involving target variables.

5.2.3 Coding of Explanatory Variables

Potential explanatory variables considered in this analysis fall into the following general categories:

- Vehicle kinematics, including impact speed (pre-crash).
- Vehicle characteristics (static).
- Driver maneuvers and attention.
- Pedestrian physiology.
- Pedestrian movement (pre-crash).

In both GIDAS and PCDS, additional variables were computed from existing ones: squared impact speed of the vehicle (to account for possible non-linearities between impact speed and injury severity), kinetic energy of the vehicle, and body mass index (BMI) of the pedestrian. Ratios between anthropometric values and vehicle dimensions were also constructed. Since only basic vehicle profile characteristics were coded in GIDAS, additional data sources were used to reconstruct some of these characteristics for the purpose of analysis (see Sect. 5.2.4). The notation "GIDAS" and "PCDS" used for labeling the variables refers to the original data set.

Tables A.1 and A.2 (see pp. 182 and 184) summarize the continuous variables used in analysis. In order to facilitate a comparative interpretation of odds ratios from different factors, normalizing transformations were applied to the remaining factors: Continuous variables, with the exception of impact speed in PCDS, were transformed by subtracting the mean and dividing by the standard deviation (SD) computed from the full sample (i.e., including cases with missing impact speed or with pedestrian age <4). Impact speed in PCDS was scaled by the mean, i.e., divided by the mean. The most important variables were tested to ensure that this procedure did not result in a significant change in the mean of those variables compared with the reduced data set.

Following standard procedures of data preparation for logistic regression [28], non-ordinal categorical variables were recoded considering each category as a separate binary. For example, for "attempted avoidance maneuvers of the car" in PCDS, the combinations of steering, braking, and accelerating were coded separately as new binary variables. Ordinal (non-continuous) variables were recoded as a cumulative binary sequence. Variables were defined to distinguish between different groups of automobiles and light trucks (see Tables A.3, p. 186, and A.4, p. 189, for a complete list of variables).

5.2.4 Treatment of Missing Data

Each case must have a complete set of valid data to be suitable for logistic regression. In case of missing data, the simplest procedure, known as list-wise deletion, is to exclude all cases for which even a single explanatory variable is missing. For example, of the 915 cases in the GIDAS data set, the percentages of missing values were

- 21.7 % for pedestrian body height,
- 21.6 % for pedestrian body weight, and
- 9.8 % for vehicle crash weight.

List-wise deletion would have the effect that the three variables mentioned above result in a loss of 30.3 % of all cases.

In this thesis, imputation of missing values was carried out and additional data were utilized for the analysis (using secondary data resources as far as available) for both data sets. Imputation narrows the resulting distributions of explanatory variables and initially leads to an underestimate of the variance in logistic regression. Nonetheless, imputation is generally thought to be preferable to the simpler procedure of list-wise deletion, both by avoiding a loss of statistical power and by minimizing biases.

Moreover, the variance underestimation process can be appraised by several methods: One method for quantifying a possible variance underestimate due to imputation is to generate additional "virtual missing data" equal to the original fraction of missing data (i.e., doubling the percentage) and then evaluate the resulting additional variance of regression coefficients resulting from this additional missing data; this procedure will be utilized below (see Sect. 5.3.2).

For the statistical questions of interest here, it was useful to carry out two preliminary steps of data preparation: imputation and augmentation.

The variables describing pedestrian body height and weight as well as the crash weight of the vehicle (i.e., weight at time of first impact) were missing for a considerable number of cases. Pedestrian body height and weight were imputed using anthropometric data from the Statistisches Bundesamt (Federal Statistical Office) for ages 18 and older [29]. For children and adolescents (ages 4–17), data from a survey on health were used [30]. To this end, body height and weight were imputed using the mean for each group defined by given sex and age. As a result of imputing those two variables, 99.7 % of the 915 cases were available for analysis without crash weight.

The crash weight of the vehicle was imputed by taking the unladen weight of the vehicle (as given by the registration papers) and adding the mean of the difference of the crash weight and the unladen weight derived from all cases where both variables are coded. As a result, 94.3 % of the 915 cases are available for analysis including crash weight. In total, 861 cases (94.1 %) are available in GIDAS after imputing all three variables mentioned above.

In the augmentation step, secondary data sources describing the vehicle profile were fused with the GIDAS data set, as the coded information is very sparse, as not many variables describing vehicle profile characteristics are included in GIDAS. Those characteristics are supposed to influence injury severity and thus should be included as explanatory variables in the analysis. Fusion was performed by referencing the coded vehicle make and model (37.0 % of the cases do not have this information coded) and associating this coded information with profile characteristic data. Missing information for vehicle make and model was added by single-file analysis (using pictures of the vehicle and the detailed accident file). During this process, many other variables were checked for consistency and were added if missing or corrected if miscoded in the data set to ensure a maximum of data quality for analysis. Except for one case, every vehicle model was added to the data.

The geometric quantities of 219 vehicles had previously been generated within another GIDAS related project; the report published included detailed instructions for measurement [31]. Data for 20 additional vehicles were generated using 3D-models available at BMW Group. In addition to the values described in the report, characteristic angles (e.g. around the bonnet leading edge or at the windshield) as well as wrapping distances were calculated from the given data points. An overview of all measurements is given in Figs. A.1–A.3, pp. 179–180. In total, 821 cases are available with all geometric values coded.

Important geometric quantities for the PCDS data set are given Figs. A.4, p. 181, and A.5, p. 181. Corresponding measurements for the pedestrian are illustrated in Fig. A.6, p. 182, both for GIDAS and PCDS.

In the PCDS data set, only pedestrian body height and weight have a substantial number of missing cases (9.5–10.0 % respectively). The body height of shoulder, hip, and knee also have about 13 % missing values each. The missing percentage due to list-wise deletion is 14.4 %. Body height and weight were imputed using anthropometric data from the National Health and Nutrition Examination Survey (NHANES) [32] standardized by sex and age; in order to take account for the study years 1994–1998, the average of the NHANES III (1988–1994) and NHANES (1999–2002) studies was used for imputation. The shoulder, hip, and knee height were imputed by calculating the ratio to body height for each sex and age group separately and multiplying that ratio by the known or imputed body height; the ratios had very narrow distributions within these groups. The full 369 cases were available for analysis after imputation.

Impact speed was not imputed in either data set as it is the most important predictor for injury causation as mentioned above.

5.2.5 Statistical Models and Methods

The statistical procedures used in this thesis are summarized in this subsection together with hints for further literature. PASW Statistics 18 and Microsoft Office Excel were used for computations. T-tests and Mann-Whitney tests were performed to assess internal relationships between binary and continuous explanatory variables, particularly vehicle impact speed. Pearson and Spearman correlations were used to assess possible correlations among continuous variables.

The t-test is a parametric method for comparing two mean values, e.g., for the difference in means between two groups or one mean value with an expected value [33, 34]. T-tests require a random sample, normally distributed and metric raw data, and homogeneous variances [33–35].

The Mann-Whitney test is a non-parametric rank test, which compares two independent samples [34, 35]. The Mann-Whitney test is used here, if some prerequisites for the t-test, e.g., the homogeneity of variances, are not given.

Possible correlations between continuous variables and vehicle impact speed were tested using Pearson and Spearman correlations. The Pearson correlation can find a correlation between variables independent of their scaling [34, 35]. Prerequisite are two continuous variables [34]. Spearman correlation uses the Bravais-Pearson correlation coefficient applied to ranks [34, 36]. As a consequence, it is also applicable to ordinal data [36].

A binary logistic regression model estimates the effect of one or several factors on the probability of a defined binary outcome [37]. The estimate can be interpreted as a group membership or the risk associated with the explanatory factors contained in the model [37, 38]. The explanatory factors can be continuous, discrete or dichotomous [38].

In the binary logistic regression approach, the estimated probability p_i for a pedestrian injury to reach or exceed the severity level in question is obtained from a given model via the formula

$$p_i = \frac{\exp\left(\beta_0 + \beta_1 x_{1,i} + \cdots + \beta_k x_{k,i}\right)}{1 + \exp\left(\beta_0 + \beta_1 x_{1,i} + \cdots + \beta_k x_{k,i}\right)} \tag{5.1}$$

where $x_{1,i}, \ldots, x_{k,i}$ are explanatory factors for the collision such as impact speed, vehicle dimensions, etc., and $\beta_1 \ldots \beta_k$ are model coefficients which are estimated in the regression process by the well-known maximum likelihood method [37, 38]. The likelihood of a model is proportional to the probability of observing the data, given the values of the model parameters. *Maximum* likelihood is a search for the parameters that maximize this probability.

A common description of the logistic formula, Eq. 5.1, uses the logit transformation:

$$\text{logit}\,(p_i) = \ln\left(\frac{p_i}{1 - p_i}\right) = \beta_0 + \beta_1 x_{1,i} + \cdots + \beta_k x_{k,i} \tag{5.2}$$

The probability estimate obtained from a model of the form of Eq. 5.1 may also be thought of as a "risk score". This probability provides an estimate of the proportion of occurrence and non-occurrence [39]. The odds is the probability of occurrence relative to probability of non-occurrence [39]. The odds are defined as [39]:

$$odds = \frac{p_i}{1 - p_i}. \tag{5.3}$$

Table 5.2 Generic 2×2 contingency table [37]

Test	Outcome	
	1	0
1	a	b
0	c	d

Following directly, probability and odds are connected via the following formula [39]:

$$p_i = \frac{\text{odds}}{1 + \text{odds}} \qquad (5.4)$$

Probability, odds, and logit are different ways for expressing the same information [40].

In a 2×2 contingency table (Table 5.2), the definition of an odds ratio is equivalent to

$$OR = \frac{ad}{bc}. \qquad (5.5)$$

If only one factor, say x_1, is entered into binary regression, the regression is referred to as "univariate" and the quantity $\exp(\beta_1)$ then known as the "crude" or "unadjusted" odds ratio for the factor x_1. The quantity $\exp(\beta_j)$ resulting from multivariate regression is known as the "adjusted odds ratio" of the explanatory factor with the label j for the outcome in question [37]. Further information on logistic regression can be found in the literature, e.g., [28, 37, 38, 40, 41]. A practical example of the calculation of p_i is given later on using actual results (see p. 125).

Univariate and multivariate binary logistic regression is used to determine unadjusted and adjusted odds ratios (respectively) and to construct risk scores for binary endpoints MAIS2+, MAIS3+, MAIS4+, MAIS5+, ISS9+, ISS16+, ISS25+, and for fatalities. Each collision is considered as a statistical unit. A minimum of 25 cases per group (e.g., minimum 25 cases with ISS9+ and minimum 25 cases with an ISS <9) is taken as the requirement for multivariate logistic regression in this analysis.

Variables with suspected impact on injury severity were first tested for univariate impact; multivariate logistic regression models were constructed for the subgroups mentioned above and evaluated for the binary injury endpoints of interest. Model selection in the multivariate models was performed by standard forward elimination using the likelihood ratio statistic. Factors that fail to be significant in a particular multivariate model are regarded as associated with a β coefficient of zero or equivalently with an odds ratio of one. Failure to reach significance in this context does not necessarily mean that a factor is truly irrelevant, but simply that it is not possible to reject the null hypothesis at the assumed level of significance. The 95 % confidence intervals of odds ratios give an indication about the validity of the findings. In the case of an odds ratio, a significant p-value ($p \leq 0.05$) is equivalent to the statement that the 95 % confidence region does not include the value one. Further explanations on the practical interpretation of odds ratios are given in Sect. 5.3.

One basic problem is to select one of a number of given models of different dimensions [42]. The maximum likelihood would lead to the selection of the model with the highest dimensionality [42]. The Akaike and Bayes information criteria allow an assessment of model fit that includes parsimony adjustment [38].

Following [43], AIC and BIC are defined as:

$$AIC = -2 \cdot LL + 2 \cdot (k + 1) \tag{5.6}$$

$$BIC = -2 \cdot LL + ln\,(n) \cdot (k + 1) \tag{5.7}$$

In Eqs. 5.6 and 5.7, LL is the log-likelihood, k the number of model parameters, and n the number of cases. Lower values of both AIC and BIC indicate improved model fit [38]. However, they both lack a normalized scale, so "low values" have to be seen in relation to models in comparison [38]. These relative differences in AIC and BIC are useful in ranking models with respect to predictive quality despite different numbers of model parameters. Further indications on relative differences in BIC and their meaning for variable selection is included in [39]. The BIC is clearly related to AIC, but it has a stronger emphasis on parsimony or over-fitting penalty.

The area under the curve (AUC) of the receiver operating characteristics (ROC) will also be evaluated as indicator for both *in-sample* and expected *out-of-sample* model quality. ROC comes from the context of electronic signal detection and is a plot of sensitivity versus specificity for a variety of cut-off points. A cut-off point defines the decision boundary, e.g., of a risk score, between the binary classification "one" (injury or fatality predicted to occur) versus "zero" (injury or fatality predicted not to occur). The optimization criterion is AUC: the larger AUC, the better the discriminatory performance of the model, independently of the cut-off point. Theoretically, if the AUC of a statistical model is 1.0, it is a perfect predictor; if the ROC AUC equals 0.5, the factors have no meaning at all (i.e., the result is random) [37]. Applied to logistic regression, the ROC AUC measures overall quality [37, 38]. Kleinbaum and Klein [37] suggests grading guidelines for AUC values in 0.1 steps: failed (0.5–0.6), poor (0.6–0.7), fair (0.7–0.8), good (0.8–0.9), and excellent discrimination (0.9–1.0).

Cross-validation was used to evaluate the expected *out-of-sample* predictive accuracy as well as the statistical question of robustness of the models or stability of the regression estimate [35]. It is also the preferable procedure if no additional data for validation are available. This procedure represents an important step toward an estimation of a realistic out-of-sample predictive power. The underlying question addresses the general validity and reliability of statements based on models derived from limited data sets. In the specific case of 10-fold cross-validation used here [44], the logistic regression model is repeatedly evaluated using nine-tenths of the data for training and one-tenth for assessment. The correlation between the predicted and real sample is a measure of stability [35]; large discrepancies indicate over-fitting and lack of generalizability [38].

ROC AUC was found to be an appropriate statistic to quantify the amount of optimism in the models. The optimism, defined here as the difference of the statistic for the full-data model and for the mean of the 10 cross-validation models, estimates the loss of accuracy, if the models are used to predict data not included in the data set used during training. A small value of optimism is an indicator for better performance in the field. A large value of optimism is usually an indicator for over-fitting.

False-classification rate with a risk-adjusted cut-off was also considered as a statistic for the cross-validation. The cut-off value corresponds to the fraction of cases in the outcome category. If the predicted risk is greater than the cut-off value, the case is classified as 1 (else 0). The false-classification rate seems to be less suited as a statistic for this study as it is highly sensitive to the low number of cases in each test group and therefore produces high variance.

Issues concerning confounding factors and multicollinearity, which commonly occur in observational data sets, will be addressed as they arise. Multicollinearity in the present context refers to the fact that within a multivariate model, β regression coefficients of correlated explanatory factors are interdependent. Thus, the apparent predictive impact of one factor can depend on whether or not a distinct but correlated factor is included in the analysis or attains significance. The implications and interpretation are included in Sect. 5.3.

5.2.6 Verifying Plausibility of Injury Probability Models

5.2.6.1 Definition and Testing of Constraints

Injury probability models can be used in two different ways:

1. Prediction of one specific level of injury (e.g., ISS16+).
2. Prediction of a variety of different cumulative injury levels (e.g., ISS0-8, ISS9-15, ...).

In the first case, the models given are ready to use and provide the highest possible explanatory value with respect to the training data. In the second case, the models have to fulfill additional plausibility criteria in order to deliver reasonable results.

Two constraints seem meaningful and reasonable:

1. If the collision speed equals zero ($v_c = 0$ kph), the injury probability is defined as zero ($p = 0$).
2. Given the sets A, B, C, and D with the relationship $D \subseteq C \subseteq B \subseteq A$ (see Fig. 5.1), the probabilities for the different sets must follow:

$$p_A \leq p_B \leq p_C \leq p_D. \tag{5.8}$$

As an example, this means for injury categories based on ISS: ISS25+ is a subset of ISS16+ which is a subset of ISS9+ ($ISS25+ \subseteq ISS16+ \subseteq ISS9+$). The probabilities for each ISS group *must* satisfy the following relationship $p_{ISS25+} \leq p_{ISS16+} \leq p_{ISS9+}$.

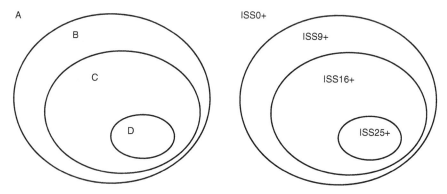

Fig. 5.1 Visualization of sets; abstract and with ISS values

Constraint 1 can be seen as reasonable addition to the probability models [45]. This can result in a point of discontinuity at $v_c = 0$ kph for some models.

Constraint 2 has to be regarded as mandatory. First of all, each set of models has to be tested, if the relationship 5.8 is fulfilled for all possible parameter combinations of significant explanatory factors. If the constraint 5.8 has not been taken into account explicitly in model development, then it needs to be tested. To this end, a conditional probability simulation can be used which generates synthetic vehicle-to-pedestrian accidents with all input parameters necessary for the models in question. The objective of the simulation is to generate the conditional probabilities for injury levels for all combinations of explanatory parameters, even rarely occurring parameter combinations.

The conditional probability simulation generates a virtual sample of 100,000 frontal vehicle-to-pedestrian accidents. The required parameters are drawn randomly from realistic distributions using Monte-Carlo techniques. Pedestrian attributes are taken from the official statistics, and vehicle attributes from the data set, both described in Sect. 5.2.4. The vehicle impact speed is drawn from a uniform distribution containing speeds up to 80 kph. (Please note that this conditional probability simulation is completely distinct from the simulation described in Sect. 3.4.)

A possible violation of constraint 2 has several causes. The binary logistic models can deliver extreme values when one or more factors entered have extreme values. In other words, the prediction on the boundaries of the models can lead to implausible results with respect to constraint 2. The reason for this is not the multiplicity of factors within the models (compared to univariate models), but the "proximity" of outcome variables.

This phenomenon is *always* present in combination with probability models described. The reason it did not yet appear in the literature is that the published models either focus on one outcome variable only [5, 6] or the outcome variables are rather far apart (e.g., MAIS2+ and MAIS5+ in [4]).

5.2.6.2 Approaches for Correct Implementation

In case only one outcome variable (e.g., ISS16+) is of interest, constraint 2 is inapplicable. However, if more injury severity levels should be assessed at once, the constraint must be fulfilled. To this end, several approaches are considered here. One simple approach in the case of a single explanatory variable is to omit the regression constant β_0. This automatically fulfills constraint 2, but leads to an unreasonable conclusion: The constant has, for example, the effect that even at very low speeds the injury probability does not equal zero. With respect to the data, this is a meaningful and reasonable result, as there are several reported accidents with very low impact speeds, but considerable levels of injury severity. Since this simple approach raises problematic questions, another approach is introduced in the following.

A new set of models using the conditional probability identity is constructed:

$$p(A \cap B) = p(A|B) \cdot p(B) \qquad (5.9)$$

Figure 5.1 visualizes the different sets in both abstract form and as example for ISS. Translated to ISS, this would mean, for example, that the (unknown) probability $p_{ISS16+} = p(\text{ISS16+}|X)$ (X stands for the different factors in logistic regression) is calculated:

$$p(\text{ISS16+}|X) = p(\text{ISS16+}|ISS9+, X) \cdot p(\text{ISS9+}|X) \qquad (5.10)$$

For example, the multivariate model delivering $p(\text{ISS9+}|X)$ is known from Eq. 5.17, p. 120, (for GIDAS data), whereas $p(\text{ISS16+}|ISS9+, X)$ has to be developed. The way of constructing $p(\text{ISS16+}|ISS9+, X)$ is identical to the way Eq. 5.18, p. 120, was constructed, with the difference that all operations are performed on the ISS9+ subset of the data.

In general, different approaches based on conditional probabilities are possible, depending on the research question and the quantity of outcome variables (i.e., different sets). Figure 5.2 gives all possible ways for constructing injury probability models using conditional probabilities (the more outcome variables, the more opportunities for construction), in this example for ISS as outcome variable.

Option *a* uses the $p(ISS9+X)$ model (for example, given by Eqs. 5.17, p. 120, and 5.28, p. 127, below). The other probabilities are calculated as (the index gives reference to the option):

$$p_a(\text{ISS16+}|X) = p(\text{ISS16+}|ISS9+, X) \cdot p(\text{ISS9+}X) \qquad (5.11)$$
$$p_a(\text{ISS25+}|X) = p(\text{ISS25+}|ISS16+, X) \cdot p_a(\text{ISS16+}|X) \qquad (5.12)$$

Option *b* uses the $p(ISS25 + |X)$ model (for example, given by Eqs. 5.19, p. 120, and 5.30, p. 120, below). The other probabilities are calculated as:

Fig. 5.2 Different
approaches to construct
plausible injury probability
models based on conditional
probability identity

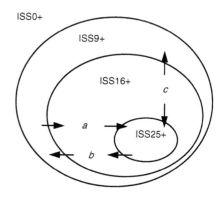

$$p_b(\text{ISS16+}|X) = p(\text{ISS16+}|ISS24-, X) \cdot p(\text{ISS24-}|X) + p(ISS25+|X)$$
$$= p(\text{ISS16+}|ISS24-, X) \cdot (1 - p(\text{ISS25+}|X)) + p(ISS25+|X)$$
$$\tag{5.13}$$

$$p_b(\text{ISS9+}|X) = p(\text{ISS9+}|ISS15-, X) \cdot p(\text{ISS15-}|X) + p_b(ISS16+|X)$$
$$= p(\text{ISS9+}|ISS15-, X) \cdot (1 - p_b(\text{ISS16+}|X)) + p_b(ISS16+|X)$$
$$\tag{5.14}$$

Option c uses the $p(ISS16 + |X)$ model (for example, given by Eqs. 5.18, p. 120, and 5.29, p. 127, below). The other probabilities are calculated as:

$$p_c(\text{ISS9+}|X) = p(\text{ISS9+}|ISS15-, X) \cdot p(\text{ISS15-}|X) + p(ISS16 + |X)$$
$$= p(\text{ISS9+}|ISS15-, X) \cdot (1 - p(\text{ISS16+}|X)) + p(ISS16 + |X)$$
$$\tag{5.15}$$

$$p_c(\text{ISS25+}|X) = p(\text{ISS25+}|ISS16+, X) \cdot p(\text{ISS16+}|X) \tag{5.16}$$

Which option to choose depends on the research question of the current evaluation and the corresponding models used. Most intuitive are Options a and b.

5.3 Prediction of Injury and Fatality Probability

5.3.1 Univariate Models and Analysis of Potential Confounders

Univariate logistic regression was performed with respect to the explanatory variables, in part as a first step to determine which variables to enter into multivariate analysis and secondly to explore which variables are predictors of the dependent variables. As quite many factors become significant (p-value ≤ 0.05) in univariate analysis, only models for ISS and for fatalities are presented and discussed here.

The odds ratios in the tables presented below refer to an additional risk associated with a significant odds ratio >1 (or decrease associated with an odds ratio <1).

The univariate models from the GIDAS data set are summarized in Tables 5.3, 5.4 and 5.5 for the three ISS groups and in Table 5.6 for fatalities (pp. 106–109). As expected, the strongest single predictive factor for all ISS levels in question as well as for fatalities is impact speed (either as a scaled variable, squared or expressed as kinetic energy). The odds ratios associated with vehicle collision speed ($v_{c,GIDAS}$) are between 3.107 and 4.548 and refer to a hypothetical increase in collision speed equal to the scaling factor (17.0 kph). The odds ratios associated with ½ as large (8.5 kph) increase in speed would be estimated as the square root of these respective odds ratios in each case. The overriding importance of impact speed is also evident from the AIC and BIC.

Vehicle characteristics are also significant in univariate analysis. Increasing height of the rear hood opening ($HRE_{v,GIDAS}$) is associated with decreasing risk for ISS9+ (unadjusted odds ratio 0.790). The general shape of the vehicle, as classified by $type_{3,veh,GIDAS}$ is also significant. If the vehicle is classified as a van, the odds ratio >1 indicates an increase in risk (2.247 for ISS16+ and 2.808 for fatalities). This corresponds with the literature on the increase in injury or fatality risk to pedestrians associated with van-like vehicles compared to passenger cars [19, 46–51] as well as with the findings mentioned below (as higher hoods are typical for van-like vehicles) (see also Sect. 6.2). Correspondingly, a higher bumper top ($UBRL_{v,GIDAS}$) is also associated with an increase in risk (1.344 for ISS16+), whereas a longer front end lowers the risk ($HRE_{l,GIDAS}$: unadjusted odds ratio 0.759 for ISS25+ and 0.807 for ISS16+; $WUE_{l,GIDAS}$: 0.789 for ISS25+).

Age of the pedestrian (y_{ped}) is associated with a higher risk (unadjusted odds ratios between 1.863 and 2.489) [19–21]. BMI is also associated with an increase in risk (unadjusted odds ratios between 1.467 and 1.835).

Interpretation of the effects evoked by the explanatory variables on injury or mortality involves hypotheses about injury causation as well as the expected trend. Since impact speed is a dominant determinant for injury and fatality, any explanatory variable that is associated with impact speed could act as a surrogate for impact speed and thus as a confounder, i.e., be significant without having a causal relationship or mask the original effect size due to the association with impact speed. Potential associations were tested using Pearson and Spearman correlations for continuous variables and t-tests and Mann-Whitney-Tests for non-continuous (binary) variables. P-values refer to the hypotheses of a correlation (for continuous variables) or to differences between the two groups (for binary variables).

The first example for a potential correlation of two variables is body height ($h_{ped,GIDAS}$), which is associated with an *apparently* increased risk, e.g., an unadjusted odds ratio 1.474 for ISS9+ or 2.342 for fatalities. Following the findings on van-like vehicles as discussed above, *higher* vehicles impose *increased* risk, which leads to the hypothesis that increased height of the pedestrian should be beneficial and thus should lead to a decrease in risk. One possible explanation for this apparent contradiction is a significant ($p < 0.001$) correlation with impact speed (Spearman coefficient 0.219), a strong risk factor. In order to test this explanation, the

Table 5.3 Univariate results for ISS9+, age group 4+ (GIDAS)

Variable	Symbol	N	n ISS		Scaling factor	p-value	Unadjusted odds ratio	95 % CI		AIC	BIC
			0–8	9+							
Vehicle kinematics											
Impact speed	$v_{c,GIDAS}$	877	682	195	17.0	<0.001	3.622	2.889	4.541	754	764
Impact speed (squared)	$v_{c,GIDAS}^2$	877	682	195	1473.5	<0.001	4.419	3.342	5.843	765	774
Kinetic energy	E_{kin}	828	640	188	1008.5	<0.001	4.297	3.204	5.763	746	755
Vehicle characteristics											
Hood rear end—vertical	$HRE_{v,GIDAS}$	787	617	170	6.9	0.050	0.790	0.624	1.000	821	830
Pedestrian physiology											
Body height	$h_{ped,GIDAS}$	874	679	195	20.3	<0.001	1.474	1.211	1.795	915	925
Age	y_{ped}	877	682	195	25.8	<0.001	1.863	1.583	2.192	875	884
Body mass index	BMI	874	679	195	5.2	<0.001	1.467	1.234	1.745	913	922
Height to upper bumper ref. l.—vert.	$r_{1,GIDAS}$	785	615	170	0.4	0.002	1.365	1.124	1.658	814	823
Height to bonnet leading edge—vert.	$r_{2,GIDAS}$	785	615	170	0.3	<0.001	1.496	1.223	1.829	808	817
Height to hood rear end—vert.	$r_{4,GIDAS}$	785	615	170	0.2	<0.001	1.670	1.345	2.072	800	809
Height to windshield up. edge—vert.	$r_{6,GIDAS}$	785	615	170	0.2	<0.001	1.603	1.293	1.986	803	813
Height to bonnet lead. edge (wrap)	$r_{3,GIDAS}$	785	615	170	0.3	<0.001	1.492	1.221	1.822	808	817
Height to hood rear end (wrap)	$r_{5,GIDAS}$	785	615	170	0.2	<0.001	1.454	1.214	1.741	807	816
Height to windshield up. edge (wrap)	$r_{7,GIDAS}$	785	615	170	0.1	<0.001	1.591	1.297	1.951	802	812
Pedestrian movement											
Walking: hazard	$hazard_{ped,GIDAS}$	877	682	195	–	0.041	0.714	0.517	0.986	929	939

Table 5.4 Univariate results for ISS16+, age group 4+ (GIDAS)

Variable	Symbol	N	n ISS 0–15	n ISS 16+	Scaling factor	p-value	Unadjusted odds ratio	95 % CI		AIC	BIC
Vehicle kinematics											
Impact speed	$v_{c,GIDAS}$	877	792	85	17.0	<0.001	4.548	3.363	6.150	410	420
Impact speed (squared)	$v_{c,GIDAS}^2$	877	792	85	1473.5	<0.001	4.405	3.243	5.985	420	429
Kinetic energy	E_{kin}	828	745	83	1008.5	<0.001	4.334	3.162	5.940	415	425
Mean braking deceleration	a_{veh}	763	689	74	33.3	0.021	0.754	0.593	0.959	485	494
Vehicle characteristics											
Body type: van-shaped	$type3,veh,GIDAS$	877	792	85	-	0.048	2.247	1.007	5.016	559	568
Upper bumper ref. l. —vert.	$UBRL_{v,GIDAS}$	787	720	67	4.0	0.015	1.344	1.060	1.705	457	466
Hood rear end—longitudinal	$HRE_{l,GIDAS}$	796	727	69	20.8	0.045	0.807	0.654	0.995	470	479
Pedestrian physiology											
Body height	$h_{ped,GIDAS}$	874	789	85	20.3	<0.001	1.838	1.338	2.527	544	554
Age	y_{ped}	877	792	85	25.8	<0.001	1.942	1.550	2.432	527	537
Body mass index	BMI	874	789	85	5.2	<0.001	1.711	1.359	2.154	541	550
Height to up. bumper ref. l. —vert.	$r_{1,GIDAS}$	785	718	67	0.4	0.023	1.404	1.048	1.881	456	466
Height to bonnet lead. edge—vert.	$r_{2,GIDAS}$	785	718	67	0.3	0.001	1.720	1.256	2.356	449	458
Height to hood rear end—vert.	$r_{4,GIDAS}$	785	718	67	0.2	<0.001	2.041	1.430	2.913	443	452
Height to winds. up. edge—vert.	$r_{6,GIDAS}$	785	718	67	0.2	<0.001	1.884	1.327	2.676	447	456
Height to bonnet lead. edge (wrap)	$r_{3,GIDAS}$	785	718	67	0.3	0.002	1.632	1.201	2.218	451	460
Height to hood rear end (wrap)	$r_{5,GIDAS}$	785	718	67	0.2	<0.001	1.712	1.348	2.175	443	452
Height to winds. up. edge (wrap)	$r_{7,GIDAS}$	785	718	67	0.1	<0.001	2.019	1.466	2.782	441	450

Table 5.5 Univariate results for ISS25+, age group 4+ (GIDAS)

Variable	Symbol	N	n ISS 0–24	n ISS 25+	Scaling factor	p-value	Unadjusted odds ratio	95 % CI		AIC	BIC
Vehicle kinematics											
Impact speed	$v_{c,GIDAS}$	877	826	51	17.0	<0.001	3.107	2.335	4.134	316	326
Impact speed (squared)	$v_{c,GIDAS}^2$	877	826	51	1473.5	<0.001	2.439	1.880	3.164	331	340
Kinetic energy	E_{kin}	828	779	49	1008.5	<0.001	2.437	1.843	3.223	318	327
Vehicle characteristics											
Hood rear end—longitudinal	$HRE_{l,GIDAS}$	796	752	44	20.8	0.024	0.759	0.598	0.964	340	349
Windshield up. edge—long	$WUE_{l,GIDAS}$	796	752	44	25.0	0.022	0.789	0.645	0.966	340	349
Pedestrian physiology											
Body height	$h_{ped,GIDAS}$	874	823	51	20.3	0.005	1.780	1.194	2.654	383	393
Age	y_{ped}	877	826	51	25.8	<0.001	2.116	1.585	2.825	365	375
Body mass index	BMI	874	823	51	5.2	0.001	1.618	1.224	2.139	382	391
Height to bonnet lead. edge—vert.	$r_{2,GIDAS}$	785	743	42	0.3	0.026	1.535	1.053	2.238	326	336
Height to hood rear end—vert.	$r_{4,GIDAS}$	785	743	42	0.2	0.006	1.805	1.184	2.753	323	332
Height to winds. up. edge—vert.	$r_{6,GIDAS}$	785	743	42	0.2	0.014	1.691	1.113	2.568	325	334
Height to bonnet lead. edge (wrap)	$r_{3,GIDAS}$	785	743	42	0.3	0.050	1.444	1.000	2.086	328	337
Height to hood rear end (wrap)	$r_{5,GIDAS}$	785	743	42	0.2	0.001	1.620	1.234	2.127	321	330
Height to winds. up. edge (wrap)	$r_{7,GIDAS}$	785	743	42	0.1	0.001	1.927	1.307	2.841	319	329
Pedestrian movement											
Walking: speed	$v_{ped,GIDAS}$	877	826	51	–	0.005	0.417	0.225	0.773	385	394
Walking: hazard	$hazard_{ped,GIDAS}$	877	826	51	–	0.011	0.450	0.243	0.835	386	396

Table 5.6 Univariate results for fatalities, age group 4+ (GIDAS)

Variable	Symbol	N	n not fatal	n fatal	Scaling factor	p-value	Unadjusted odds ratio	95 % CI		AIC	BIC
Vehicle kinematics											
Impact speed	$v_{c,GIDAS}$	915	866	49	17.0	<0.001	3.995	2.939	5.429	271	280
Impact speed (squared)	$v_{c,GIDAS}^2$	915	866	49	1473.5	<0.001	3.024	2.320	3.941	281	291
Kinetic energy	E_{kin}	863	817	46	1008.5	<0.001	3.111	2.326	4.161	264	274
Mean braking deceleration	a_{veh}	797	752	45	33.3	0.012	0.674	0.496	0.918	344	353
Vehicle characteristics											
Body type: van-shaped	$type_{3,veh,GIDAS}$	915	866	49	–	0.026	2.808	1.130	6.975	382	392
Pedestrian physiology											
Body height	$h_{ped,GIDAS}$	912	863	49	20.3	<0.001	2.342	1.482	3.699	368	378
Age	y_{ped}	915	866	49	25.8	<0.001	2.489	1.827	3.390	348	358
Body mass index	BMI	912	863	49	5.2	<0.001	1.835	1.392	2.419	368	378
Height to up. bum. ref. l. —vert.	$r_{1,GIDAS}$	819	775	44	0.4	0.025	1.508	1.052	2.162	341	351
Height to bonnet lead. e. —vert.	$r_{2,GIDAS}$	819	775	44	0.3	0.041	1.456	1.016	2.086	342	352
Height to hood rear end —vert.	$r_{4,GIDAS}$	819	775	44	0.2	0.010	1.684	1.131	2.508	339	349
Height to winds. up. edge —vert.	$r_{6,GIDAS}$	819	775	44	0.2	0.015	1.644	1.102	2.452	340	349
Height to bonnet lead. e. (wrap)	$r_{3,GIDAS}$	819	775	44	0.3	0.049	1.427	1.001	2.034	343	352
Height to hood rear end (wrap)	$r_{5,GIDAS}$	819	775	44	0.2	0.001	1.557	1.204	2.013	337	347
Height to winds. up. edge (wrap)	$r_{7,GIDAS}$	819	775	44	0.1	0.001	1.941	1.335	2.820	333	343
Pedestrian movement											
Walking: speed	$v_{ped,GIDAS}$	915	866	49	–	0.001	0.318	0.164	0.619	373	383
Walking: hazard	$hazard_{ped,GIDAS}$	915	866	49	–	0.042	0.529	0.287	0.976	382	391

magnitude of the influence of impact speed contained in body height was quantified by a linear regression. Linear regression analysis would imply that this correlation contributes to the effect described above, but would not explain it entirely. However, body height seems to be quite strongly associated with crashes involving high speeds: The correlation below 50 kph still exists ($p < 0.001$), but is a bit weaker (Spearman coefficient 0.158). The remaining effect size not explained by a correlation with impact speed could also have been attributable to selection effects and correlations with other factors in the data set or characteristics which have not been observed and coded. Summarizing, the unadjusted odds ratios (>1) for ratios between body height and vehicle characteristics are all counterintuitive; due to the correlations involved, multivariate analysis is essential for proper interpretation of these causal factors. The explanatory value of those factors is again discussed for multivariate models in Sect. 5.3.2.

Another example of a potential confounder is acceleration of the vehicle (a_{veh}), as given in Tables 5.4 and 5.6 with unadjusted odds ratios of 0.754 and 0.674. There could be a causal connection between acceleration and injury causation, as the vehicle pitches and the impact kinematics change. Since the acceleration before impact is coded as mean value determined during reconstruction, the actual value at the moment of impact is unknown (as well as the actual size of the pitch). In addition, a_{veh} is weakly but significantly correlated ($p = 0.004$) with impact speed (Spearman coefficient -0.102).

The speed of the pedestrian is coded in two variables $v_{ped,GIDAS}$ and $hazard_{ped,GIDAS}$. The original information for pedestrian speed was recoded into those variables following these hypotheses: First, it could make a difference for the genesis of an accident (and the perception and interpretation of the situation by the driver) if the pedestrian is moving slow or fast. Second, a standing pedestrian or fast moving pedestrian might be harder to detect for the driver. In the first case, because he is not moving or, in the second case, because he is approaching the situation fast and thus leaving the driver less time to react. This behavior is therefore coded as more hazardous. These hypotheses seem to make sense, as both variables are correlated with impact speed. Pedestrians with a low speed ($v_{ped,GIDAS}$) have a significantly lower average impact speed than the faster ones in the data set (28.4 kph vs. 30.4 kph, Mann-Whitney, $p = 0.008$, $z = -2.663$). Pedestrians standing or running are defined as imposing more hazard to the situation and are associated with a higher average collision speed (30.6 kph vs. 28.3 kph, $p = 0.015$, Mann-Whitney, $z = -2.432$).

Nevertheless, both variables are associated with an unexpected unadjusted odds ratio (between 0.318 and 0.714) in Tables 5.3, 5.5 and 5.6, as the groups with significantly faster impact speed are associated with a decrease in risk. The reason for that is a dependency on other factors that influence injury causation. There are significant ($p < 0.001$) differences regarding, e.g., pedestrian age or BMI between the groups for both variables. The mean pedestrian age is about 17 years higher for the low hazard group ($hazard_{ped,GIDAS}$) and about 19 years higher for the low speed group ($v_{ped,GIDAS}$). The mean BMI is about 2.5 points higher in those groups. It is important to keep in mind that the assignment of categories, such as running or walking in GIDAS, involves self-reported and subjective perceptions as well as a posteriori estimations. These assignments could have large correlations to important causal factors, such as speed or age. Moreover, in view of the non-monotonic relationship between pedestrian speed and age, the correlation analysis may not capture all possible mechanisms for confounding.

Another interesting finding in the data set is the connection between impact speed and sex (G_{ped}). Although not significant in the univariate results, men are subject to a greater risk in the data set than women (31.9 kph vs. 27.0 kph, $p < 0.001$, Mann-Whitney, $z = -4.177$).

The univariate results from the PCDS data set are summarized in Tables 5.7, 5.8 and 5.9 for ISS and in Table 5.10 for fatalities. The overriding importance of impact speed ($v_{c,PCDS}$) is also obvious in the PCDS data set. The unadjusted odds ratios associated with vehicle collision speed are between 8.098 and 17.225 and refer to a hypothetical increase in collision speed equal to the scaling factor (28.9 kph).

A large number of factors become significant in univariate analysis. All variables describing vehicle characteristics (except angle of front bumper $\alpha_{1,PCDS}$) are associated with a higher risk (odds ratios between 1.261 and 1.600). Growing dimensions in height, wrapping distances, or the lead of the bumper are risk factors ($\alpha_{1,PCDS}$ shows the reverse trend). A ratio of hip height to the transition point ($r_{3,PCDS}$) is also significant for ISS25+ as well as for fatalities, an increase (i.e., a taller person or a lower vehicle) imposes reduced risk (odds ratio 0.778 and 0.702 respectively).

Several variables describing the pedestrian are significant: Increasing age, weight, and BMI are associated with higher risk.

As mentioned above for the GIDAS data set, some variables in PCDS are also correlated with impact speed and are potential confounders. In this context, sex (G_{ped}) is significantly correlated with impact speed (33.0 kph for men vs. 25.5 kph for women, Mann-Whitney, $p < 0.001, z = -3.561$). The corresponding unadjusted odds ratio of 0.535 for ISS25+ and 0.503 for fatalities indicate decreased risk for female pedestrians.

A similar effect can explain the unadjusted odds ratio between 2.054 and 2.359 for evasive steering (δ_{driver}) of the vehicle (38.9 kph for evasive steering compared to 27.3 kph for others, Mann-Whitney, $p < 0.001, z = -4.534$), if one supposes that evasive steering is more likely in critical situations with higher speeds, i.e., where the driver senses that the time-to-collision is inadequate for effective mitigation by braking alone. Of the 369 accidents considered, 167 drivers attempted to brake; 53 performed an evasive maneuver (of whom 47 also braked). A similar correlation, of course, can be found for steering to the left ($\delta_{driver,l}$).

Regarding prior vehicle movements, collisions following "complex" (c_{driver}) prior maneuvers (e.g. turning, merging, lane changing, etc.) are associated with favorable unadjusted odds ratios (0.217–0.340). Again, this odds ratio can be adequately explained by the significantly (Mann-Whitney, $p < 0.001, z = -9.812$) lower mean impact speed in this category (17.6 kph compared to 35.0 kph).

Regarding the direction of the pedestrian in relation to the road, crossing pedestrians ($\omega_{ped,3}$) are hit significantly (Mann-Whitney, $p = 0.002, z = -3.095$) slower in average than pedestrians moving in other directions (28.0 kph compared to 39.9 kph),

Table 5.7 Univariate results for ISS9+, age group 4+ (PCDS)

Variable	Symbol	N	n ISS 0–8	n ISS 9+	Scaling factor	p-value	Unadjusted odds ratio	95 % CI		AIC	BIC
Vehicle kinematics											
Impact speed	$v_{c,PCDS}$	369	213	156	28.9	<0.001	17.225	9.435	31.447	340	347
Impact speed (squared)	$v_{c,PCDS}^2$	369	213	156	$(28.9)^2$	<0.001	3.928	2.825	5.464	344	352
Kinetic energy	E_{kin}	369	213	156	113.1	<0.001	37.840	15.479	92.506	341	349
Vehicle characteristics											
Front bumper lead	$x_{1,PCDS}$	368	212	156	3.0	0.001	1.428	1.148	1.776	495	502
Rear hood dist. from ground (wrap)	$w_{2,PCDS}$	369	213	156	22.2	0.021	1.291	1.040	1.604	501	509
Winds. base dist. from ground (wrap)	$w_{3,PCDS}$	369	213	156	28.9	0.029	1.274	1.025	1.584	502	510
Winds. top dist. from ground (wrap)	$w_{4,PCDS}$	369	213	156	21.1	0.020	1.309	1.043	1.643	501	509
Driver maneuvers and attention											
Avoidance: Steering left	$\delta_{driver,1}$	369	213	156	–	0.044	2.134	1.020	4.464	503	510
Avoidance: Steering	δ_{driver}	369	213	156	–	0.005	2.359	1.301	4.278	498	506
Pre-event movement car: complexity	c_{driver}	369	213	156	–	<0.001	0.340	0.212	0.543	485	493
Pedestrian physiology											
Age	y_{ped}	369	213	156	22.2	0.002	1.421	1.143	1.765	496	504
Body weight	$m_{ped,PCDS}$	369	213	156	22.6	0.033	1.269	1.019	1.581	502	510
Body mass index	BMI	369	213	156	5.7	0.027	1.269	1.028	1.566	502	509
Pedestrian movement											
Direction ped.: towards lane / crossing	$\omega_{ped,3}$	369	213	156	–	0.044	0.458	0.214	0.981	503	510

Table 5.8 Univariate results for ISS16+, age group 4+ (PCDS)

Variable	Symbol	N	n ISS 0–15	n ISS 16+	Scaling factor	p-value	Unadjusted odds ratio	95 % CI		AIC	BIC
Vehicle kinematics											
Impact speed	$v_{c,PCDS}$	369	248	121	28.9	<0.001	15.342	8.587	27.411	308	315
Impact speed (squared)	$v_{c,PCDS}^2$	369	248	121	$(28.9)^2$	<0.001	3.049	2.342	3.970	317	325
Kinetic energy	E_{kin}	369	248	121	113.1	<0.001	14.977	7.805	28.739	323	331
Vehicle characteristics											
Front bumper lead	$x_{1,PCDS}$	368	247	121	3.0	0.008	1.361	1.085	1.706	463	470
Rear hood dist. from ground (wrap)	$w_{2,PCDS}$	369	248	121	22.2	0.042	1.265	1.008	1.588	467	475
Winds. base dist. from ground (wrap)	$w_{3,PCDS}$	369	248	121	22.6	0.045	1.261	1.005	1.582	467	475
Winds. top dist. from ground (wrap)	$w_{4,PCDS}$	369	248	121	21.1	0.016	1.342	1.056	1.706	465	473
Driver maneuvers and attention											
Avoidance: Steering left	$\delta_{driver,l}$	369	248	121	–	0.013	2.539	1.221	5.278	465	473
Avoidance: Steering	δ_{driver}	369	248	121	–	0.007	2.240	1.242	4.040	464	472
Pre-event movement car: complexity	c_{driver}	369	248	121	–	<0.001	0.241	0.139	0.419	441	449
Pedestrian physiology											
Age	y_{ped}	369	248	121	22.2	0.001	1.457	1.163	1.826	460	468
Body weight	$m_{ped,PCDS}$	369	248	121	22.6	0.019	1.318	1.047	1.660	465	473
Body mass index	BMI	369	248	121	5.7	0.034	1.261	1.018	1.563	466	474

Table 5.9 Univariate results for ISS25+, age group 4+ (PCDS)

Variable	Symbol	N	n ISS		Scaling factor	p-value	Unadjusted odds ratio	95% CI		AIC	BIC
			0–24	25+							
Vehicle kinematics											
Impact speed	$v_{c,PCDS}$	369	277	92	28.9	<0.001	16.988	9.212	31.328	255	263
Impact speed (squared)	$v_{c,PCDS}^2$	369	277	92	$(28.9)^2$	<0.001	3.039	2.356	3.921	255	263
Kinetic energy	E_{kin}	369	277	92	113.1	<0.001	14.859	7.982	27.661	260	268
Vehicle characteristics											
Front bumper lead	$x_{1,PCDS}$	368	276	92	3.0	<0.001	1.600	1.235	2.074	404	412
Transition point height at contact	$h_{4,PCDS}$	369	277	92	15.6	0.034	1.278	1.018	1.603	414	422
Angle of front bumper	$\alpha_{1,PCDS}$	368	276	92	15.4	0.035	0.787	0.629	0.983	413	421
Driver maneuvers and attention											
Avoidance: Steering left	$\delta_{driver,l}$	369	277	92	–	0.035	2.235	1.056	4.726	414	422
Avoidance: Steering	δ_{driver}	369	277	92	–	0.022	2.054	1.111	3.797	413	421
Pre-event movement car: complexity	c_{driver}	369	277	92	–	<0.001	0.315	0.174	0.568	401	409
Pedestrian physiology											
Sex	G_{ped}	369	277	92	–	0.011	0.535	0.330	0.867	412	420
Age	y_{ped}	369	277	92	22.2	<0.001	1.612	1.263	2.056	403	411
Hip height to transition point height	$r_{3,PCDS}$	369	277	92	0.2	0.046	0.778	0.608	0.995	414	422
Pedestrian movement											
Direction ped.: with / against traffic	$\omega_{ped,2}$	369	277	92	–	0.056	2.769	0.976	7.860	415	423
Direction ped.: towards lane / crossing	$\omega_{ped,3}$	369	277	92	–	0.018	0.397	0.185	0.854	413	421

Table 5.10 Univariate results for fatalities, age group 4+ (PCDS)

Variable	Symbol	N	n not fatal	n fatal	Scaling factor	p-value	Unadjusted odds ratio	95 % CI		AIC	BIC
Vehicle kinematics											
Impact speed	$v_{c,PCDS}$	369	318	51	28.9	<0.001	8.098	4.789	13.693	207	215
Impact speed (squared)	$v_{c,PCDS}^2$	369	318	51	$(28.9)^2$	<0.001	1.865	1.578	2.204	215	223
Kinetic energy	E_{kin}	369	318	51	113.1	<0.001	3.725	2.528	5.491	228	236
Vehicle characteristics											
Front bumper lead	$x_{1,PCDS}$	368	317	51	3.0	0.010	1.498	1.102	2.036	293	301
Transition point height at contact	$h_{4,PCDS}$	369	318	51	15.6	0.013	1.419	1.077	1.869	294	302
Rear hood dist. from ground (wrap)	$w_{2,PCDS}$	369	318	51	22.2	0.013	1.483	1.085	2.026	294	302
Winds. base dist. from ground (wrap)	$w_{3,PCDS}$	369	318	51	22.6	0.010	1.488	1.100	2.012	294	302
Driver maneuvers and attention											
Avoidance: Steering left	$\delta_{driver,l}$	369	318	51	–	0.018	2.748	1.192	6.339	295	303
Pre-event movement car: complexity	c_{driver}	369	318	51	–	0.001	0.217	0.090	0.524	285	293
Pedestrian physiology											
Sex	G_{ped}	369	318	51	–	0.030	0.506	0.273	0.936	296	303
Age	y_{ped}	369	318	51	22.2	<0.001	1.741	1.290	2.349	287	295
Body weight	$m_{ped,PCDS}$	369	318	51	22.6	0.002	1.634	1.202	2.220	290	298
Body mass index	BMI	369	318	51	5.7	0.001	1.553	1.192	2.022	290	298
Hip height to transition point height	$r_{3,PCDS}$	369	318	51	0.2	0.027	0.702	0.513	0.960	295	303
Pedestrian movement											
Direction ped.: with/against traffic	$\omega_{ped,2}$	369	318	51	–	0.034	3.348	1.095	10.233	297	304
Direction ped.: towards lane/crossing	$\omega_{ped,3}$	369	318	51	–	0.039	0.399	0.167	0.954	297	304

which leads to favorable unadjusted odds ratios (0.397–0.458). The same significant effect shows up for pedestrians walking along the street ($\omega_{ped,2}$) indicated by unadjusted odds ratios between 2.769 and 3.348: 45.9 kph for pedestrians walking along versus 28.2 kph for others, Mann-Whitney, $p = 0.001$, $z = -3.190$.

Although not significant in univariate regression, slow-moving ($v_{ped,PCDS}$) pedestrians have a significantly lower mean impact speed than fast moving ones (26.3 kph vs. 32.0 kph, Mann-Whitney, $p < 0.001$, $z = -4.261$); pedestrian walking speed was not significantly associated with braking of the vehicle, evasive steering or speed limit. Hazardous walking behavior ($hazard_{ped,PCDS}$), i.e., fast or not moving (contrary to slow moving) is associated with an increased risk due to higher mean impact speed (32.3 kph vs. 25.6 kph, Mann-Whitney, $p < 0.001$, $z = -4.919$).

The above example highlights a fundamental difficulty with the correct interpretation of statistical results from observational (i.e., non-randomized) data sets. Since selection effects and dependencies between explanatory variables are common phenomena, a sound interpretation should include careful analysis of possible confounders and requires knowledge or hypotheses about cause-effect relations in advance based on different methods or sources of knowledge.

The univariate results presented above show that variables from each of the five categories defined in Sect. 5.2.3, describing the characteristics of the collision, the pedestrian, and the vehicle as well as the maneuvers of the vehicle and the pedestrian can be significant predictors for different ISS levels or fatalities for both data sets used. Impact speed is by far the most important predictor.

It is important to assess internal relationships between explanatory variables and impact speed, as these can distort the interpretation of the results, especially if no hypotheses about the causal connection of the variable and injury or fatality risk exist. The findings show that some variables are correlated with impact speed and those findings are interpreted. In addition, conclusions about the behavior of the driver or the hazard induced by specific maneuvers are possible by testing these variables against impact speed. This method allows for the discovery of aspects that are not coded explicitly in the data sets and thus enhance the knowledge about the genesis of the accidents as well as their course of events.

5.3.2 Multivariate Analysis: MAIS or ISS as Injury Scale

The first research question regarding the development of injury probability models is which injury scale available in the datasets should be used for the models (Sect. 5.1). The MAIS and ISS are coded in GIDAS and PCDS. Fatalities are investigated separately, as they are distributed over a wide range of MAIS or ISS values.

For the intended application of this work, it is important to verify the hypothesis that the explanatory variables could have an independent causal influence on injury severity for biomechanical or physical reasons. The variables associated with prior and evasive maneuvers of the vehicle as well as the pedestrian were omitted on the hypothesis that they have no direct causal relation, but are associated with injury due to selection effects (as described and analyzed in the preceding section). In each case, the best fitting logistic regression model (according to the AIC) is presented.

5.3.2.1 Multivariate Models Based on GIDAS

Table 5.11 gives the numbers of cases available for the following analysis in the GIDAS data set. The corresponding multivariate results are given in Table 5.12 for ISS, Table 5.13 for MAIS, and Table 5.14 for fatalities. The resulting models contain between two and six predictors. As in univariate analysis (Sect. 5.3.1), the strongest single predictive factor is impact speed (as a linear quantity or squared value, as for MAIS4+). Pedestrian age is included in every model and is positively associated with increased risk (adjusted odds ratios between 2.033 and 3.445). Pedestrian weight is included in the model for ISS16+ and is also associated with increased risk. The other variables included refer to vehicle static characteristics or ratios of body height to vehicle quantities.

Three variables describing the geometry of the vehicle are significant in multivariate analysis. The vertical measurement from the ground to the lower bumper reference line ($LBRL_{v,GIDAS}$) is associated with a decrease in risk. The longitudinal measurements ($BLE_{l,GIDAS}$ and $UBRL_{l,GIDAS}$) have the reverse effect. Both of them give indication about the shape of the vehicle front. With increased values, the probability of injury increases. The odds ratios connected with ratios of body height to vehicle geometry demand more sophisticated interpretation. As the variables in

Table 5.11 Frequencies of target variables (GIDAS)

Injury Level		Cases
	0–1	435
	2+	450
	0–2	731
MAIS	3+	154
	0–3	822
	4+	63
	0–4	856
	5+	29
	0–8	682
	9+	195
ISS	0–15	792
	16+	85
	0–24	826
	25+	51
Fatalities	Not fatal	866
	Fatal	46

Table 5.12 Multivariate logistic regression models for ISS9+, ISS16+, and ISS25+, age group 4+ (GIDAS)

Variable	Symbol	Scaling factor	p-value	Adjusted odds ratio	95 % CI	
ISS9+ (AIC: 595; BIC: 609)						
Age	y_{ped}	25.8	<0.001	2.185	1.771	2.694
Impact speed	$v_{c,GIDAS}$	17.0	<0.001	3.656	2.833	4.716
Constant	$\exp(\beta_0)$			0.192		
ISS16+ (AIC: 283; BIC: 317)						
Body weight	$m_{ped,GIDAS}$	21.3	0.002	2.209	1.351	3.612
Age	y_{ped}	25.8	<0.001	2.423	1.679	3.497
Impact speed	$v_{c,GIDAS}$	17.0	<0.001	5.097	3.506	7.410
Lower bumper reference line—vert.	$LBRL_{v,GIDAS}$	9.2	0.022	0.647	0.446	0.939
Bonnet lead. e.—l	$BLE_{l,GIDAS}$	3.0	0.033	1.419	1.030	1.955
Height to up. bum. reference line—vert.	$r_{1,GIDAS}$	0.4	0.010	0.497	0.293	0.844
Constant	$\exp(\beta_0)$			0.026		
ISS25+ (AIC: 234; BIC: 248)						
Age	y_{ped}	25.8	<0.001	2.744	1.837	4.099
Impact speed	$v_{c,GIDAS}$	17.0	<0.001	3.180	2.332	4.337
Constant	$\exp(\beta_0)$			0.022		

the ISS16+ and MAIS2+ models are associated with a decrease in risk due to an increase of the variable (i.e., higher body relative to the vehicle), the ones for the MAIS4+ and MAIS5+ model show the reverse trend. As discussed in the previous section, this effect can be attributed to a positive correlation between body height and impact speed, which influences the statistics presented here and does not allow for an interpretation of body height as an independent causal factor.

In order to quantify the variance underestimate due to imputation, multiple instances of the data set were generated as described above (Sect. 5.2.4), including an additional fraction of "virtual missing data" equal to the original fraction of missing data. To illustrate the method, the regression coefficients in the model for ISS16+ were recomputed for the two variables describing body weight and height of the pedestrian. Regarding body weight, five instances with a total of 43.2 % imputed values each have been computed, compared to 21.6 % in the original GIDAS data set. The root-mean-square RMS deviation of the regression coefficient for $m_{ped,GIDAS}$ was 6.94 %. For body height, five instances with a total of 43.4 % imputed values each have been computed, compared to 21.7 % in the original data. The RMS deviation of the regression coefficient for $r_{1,GIDAS}$ was 1.74 %. These deviations are much smaller than the confidence intervals resulting from logistic regression.

Table 5.13 Multivariate logistic regression models for MAIS2+, MAIS3+, MAIS4+, and MAIS5+, age group 4+ (GIDAS)

Variable	Symbol	Scaling factor	p-value	Adjusted odds ratio	95 % CI	
MAIS2+ (AIC: 864; BIC: 888)						
Age	y_{ped}	25.8	<0.001	2.125	1.750	2.581
Impact speed	$v_{c,GIDAS}$	17.0	<0.001	3.144	2.494	3.964
Lower bumper reference line—vert.	$LBRL_{v,GIDAS}$	9.2	0.019	0.819	0.694	0.967
Height to winds. up. edge (wrap)	$r_{7,GIDAS}$	0.1	0.013	0.782	0.644	0.950
Constant	$\exp(\beta_0)$			1.197		
MAIS3+ (AIC: 534; BIC: 553)						
Age	y_{ped}	25.8	<0.001	2.118	1.691	2.653
Impact speed	$v_{c,GIDAS}$	17.0	<0.001	2.975	2.333	3.793
Constant	$\exp(\beta_0)$			0.128		
MAIS4+ (AIC: 259; BIC: 283)						
Age	y_{ped}	25.8	<0.001	2.643	1.787	3.909
Impact speed	$v_{c,GIDAS}^2$	1473.5	<0.001	3.258	2.334	4.547
Bonnet lead. e.—l.	$BLE_{l,GIDAS}$	3.0	0.006	1.645	1.155	2.344
Height to hood rear end (wrap)	$r_{5,GIDAS}$	0.2	0.001	1.850	1.307	2.619
Constant	$\exp(\beta_0)$			0.025		
MAIS5+ (AIC: 172; BIC: 192)						
Age	y_{ped}	25.8	0.005	2.033	1.246	3.319
Impact speed	$v_{c,GIDAS}$	17.0	<0.001	2.789	1.982	3.925
Height to hood rear end (wrap)	$r_{5,GIDAS}$	0.2	0.013	1.534	1.093	2.153
Constant	$\exp(\beta_0)$			0.013		

Table 5.14 Multivariate logistic regression models for fatalities, age group 4+ (GIDAS)

Variable	Symbol	Scaling factor	p-value	Adjusted odds ratio	95 % CI	
Fatalities (AIC: 203; BIC: 218)						
Age	y_{ped}	25.8	<0.001	3.445	2.163	5.488
Impact speed	$v_{c,GIDAS}$	17.0	<0.001	3.946	2.829	5.503
Constant	$\exp(\beta_0)$			0.012		

Using Eq. 5.1, the resulting models can be written as follows:

$$P_{ISS9+,GIDAS} = \frac{1}{1 + \exp\left(1.650 - 1.296 \cdot v_{c,GIDAS} - 0.781 \cdot y_{ped}\right)} \tag{5.17}$$

$$P_{ISS16+,GIDAS} = \left(1 + \exp\left(3.631 - 0.885 \cdot y_{ped} - 0.792 \cdot m_{ped,GIDAS} - 1.629 \cdot v_{c,GIDAS} + \right.\right.$$
$$\left.\left. +0.435 \cdot LBRL_{v,GIDAS} - 0.35 \cdot BLE_{l,GIDAS} + 0.699 \cdot r_{1,GIDAS}\right)\right)^{-1} \tag{5.18}$$

$$P_{ISS25+,GIDAS} = \frac{1}{1 + \exp\left(3.822 - 1.157 \cdot v_{c,GIDAS} - 1.009 \cdot y_{ped}\right)} \tag{5.19}$$

$$P_{Fatalities,GIDAS} = \frac{1}{1 + \exp\left(4.391 - 1.373 \cdot v_{c,GIDAS} - 1.237 \cdot y_{ped}\right)} \tag{5.20}$$

An example of the practical use of the models is given using Eq. 5.17. Note that the parameters have been transformed (see Sect. 5.2.3) using mean and standard deviation. For example, a vehicle impact speed of 35 kph and a pedestrian age of 20 years are used. Together with mean and standard deviation from Table A.1, p. 182, Eq. 5.17 delivers the following probability:

$$P_{ISS9+,GIDAS,expl.} = \frac{1}{1 + \exp\left(-\ln(0.192) - \ln(3.656) \cdot v_{c,GIDAS} - \ln(2.185) \cdot y_{ped}\right)}$$

$$= \frac{1}{1 + \exp\left(1.650 - 1.296 \cdot v_{c,GIDAS} - 0.781 \cdot y_{ped}\right)}$$

$$= \frac{1}{1 + \exp\left(1.650 - 1.296 \cdot \left(\frac{35 - 29.35}{17.04}\right) - 0.781 \cdot \left(\frac{20 - 35.91}{25.83}\right)\right)}$$

$$= 0.155$$

The predictive performance of the models is quantified using receiver operator characteristic (ROC) analysis. The *in-sample* predictive accuracy is given by the area under the curve (AUC) of the ROC. Compared with the results of the 10-fold cross-validation, the expected *out-of-sample* accuracy can be estimated. Table 5.15 gives the corresponding results. The predictive quality, in-sample and out-of-sample, is remarkably high (ROC AUC 0.749–0.915). The optimism, quantifying the difference between in-sample and out-of-sample predictive accuracy, is relatively small for all models (except for MAIS5+). The model for fatalities has the highest expected out-of-sample accuracy with 0.915.

The question whether MAIS or ISS should be used as target variable for injury or mortality probability models can be addressed for the GIDAS data set using the expected out-of-sample performance of the models. The ISS-based models have a mean ROC AUC between 0.827 and 0.879, whereas the MAIS-based ones have 0.749–0.873. The models are not directly comparable, as they predict probabilities of different injury scales. The ISS-based models tend to have a higher accuracy, as was suspected from the advantages of ISS over MAIS as stated in the medical literature (see Sect. 5.1).

Table 5.15 Predictive accuracy of the models given by Eqs. 5.17–5.20 (GIDAS)

Model	Full-data model				Cross-validation			
	ROC AUC	95 % CI		k	ROC AUC	SD	SE	Optimism
ISS9+	0.831	0.798	0.864	2	0.827	0.046	0.015	0.004
ISS16+	0.921	0.891	0.950	6	0.879	0.070	0.023	0.041
ISS25+	0.883	0.841	0.925	2	0.861	0.065	0.022	0.022
MAIS2+	0.764	0.732	0.797	4	0.749	0.041	0.014	0.015
MAIS3+	0.818	0.777	0.860	3	0.811	0.080	0.027	0.008
MAIS4+	0.904	0.869	0.938	4	0.873	0.030	0.010	0.031
MAIS5+	0.888	0.841	0.935	3	0.759	0.171	0.057	0.129
Fatalities	0.925	0.891	0.959	2	0.915	0.054	0.018	0.010

The number of included parameters is given by k

The relatively high standard deviation (SD) as well as standard error of the mean (SE) obtained in cross-validation seem to decrease with higher case numbers and become smaller than the optimism.

5.3.2.2 Comparison to Existing models

There are a few models in the literature which are based on GIDAS and predict the probability for a particular injury level or for fatalities. The coefficients of the models are not directly comparable, as each model uses a different data set and probably a different scaling for the explanatory factors. The first two models for MAIS2+ and MAIS5+ are based on impact speed. They are not included explicitly in the publication, but only given as diagram [4].

Another model derived from the GIDAS data for MAIS2+ is given in [5]:

$$p_{MAIS2+} = \frac{1}{1 + \exp\left(2.54 - 0.06 \cdot v_{c,GIDAS} - 0.02 \cdot y_{ped}\right)} \tag{5.21}$$

This model includes impact speed and pedestrian age. Geometric quantities are not included (as given in Table 5.13). The ROC AUC of model 5.21 with respect to the data used for this thesis is 0.758. This ROC AUC is comparable to the corresponding value given in Table 5.15.

Regarding fatalities, the following model also includes impact speed and age as independent predictors (as the one presented in Table 5.14). It is again based on GIDAS and valid for pedestrians at age 15 and older [6]:

$$p_{fatal} = \frac{1}{1 + \exp\left(9.1 - 0.095 \cdot v_{c,GIDAS} - 0.040 \cdot y_{ped}\right)} \tag{5.22}$$

The ROC AUC of 0.898 for pedestrians 15 and older is identical with the ROC AUC for the model presented in Table 5.14 within the same age group.

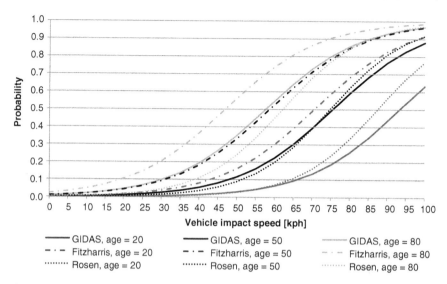

Fig. 5.3 Mortality estimates for different pedestrian ages depending on vehicle impact speed, as given by Eqs. 5.20, 5.22, 5.23

Fitzharris and Fildes constructed another model for mortality prediction [52], which is unpublished but included in [3]. No further information, e.g., on data used, sampling procedures or boundary conditions, is available. The model uses vehicle impact speed and pedestrian age ([52] following [3]):

$$p_{fatal} = \frac{1}{1 + \exp\left(6.302424 - 0.080358 \cdot v_c - 0.03166 \cdot y_{ped}\right)} \qquad (5.23)$$

The ROC AUC is 0.923 with respect to the data used for this thesis, which is comparable to the ROC AUC of 0.925 given in Table 5.15 for the corresponding model.

Figure 5.3 gives the mortality estimates of the three models (Eqs. 5.20, 5.22, and 5.23) as function of vehicle impact speed. The model constructed in this thesis as well as the model by Rosen give rather similar estimates. The model by Fitzharris tends to higher probabilities for given values of impact speed and pedestrian age. As both models are constructed using the GIDAS data set and include the same parameters, this is a plausible and expected finding. As discussed above, hardly any information besides the formula itself [3] is published for the Fitzharris model, consequently, no further interpretation of the different model estimates is possible.

The review of models available in the literature shows that ISS as outcome variable has been considered [52], but no model parameters have been published so far [3]. Explicit models for MAIS other than MAIS2+ have not been published. Regarding fatalities, a comparable model was derived in a previous study [6] and is confirmed by the results of this study.

5.3.2.3 Multivariate Models Based on PCDS

For the PCDS data set Table 5.16 gives the numbers of cases available. The corresponding multivariate logistic regression models are given in Table 5.17 for ISS, Table 5.18 for MAIS, and Table 5.19 for fatalities. The resulting models contain between two and four predictors. Impact speed is the most important predictor and is included in every model. Note that impact speed is scaled differently than in the GIDAS data set, as it is scaled by the mean only.

Pedestrian age is also included in every model except MAIS5+ and is associated with an increased risk (odds ratio 1.669–2.163). BMI is included in the model for fatalities and is associated with increased risk (odds ratio 1.742).

Different variables describing the geometry of the vehicle as well as the pedestrian stature are significant in multivariate analysis and included in every model (except MAIS2+). Increasing hip height ($h_{h,ped,PCDS}$) is associated with decreased risk (see Table 5.19). Ratios of body height to vehicle geometry are included in all ISS as well as in the MAIS4+ model and are also associated with decreased risk. The vehicle characteristics significant in the MAIS3+ and MAIS5+ models support these findings: Increased height of the vehicle front is associated with higher risk. The well-known risk imposed by light trucks or vans which typically have those geometric quantities are in line with these findings (see above).

It is important to interpret the results with respect to multicollinearity (correlations between factors within the PCDS data set): In the ISS25+ model, the ratio of hip height to the front-top transition point ($r_{3,PCDS}$) is associated with a small ($0.298 < 1$)

Table 5.16 Frequencies of target variables (PCDS)	Injury Level		Cases
		0–1	157
		2+	211
		0–2	226
	MAIS	3+	142
		0–3	288
		4+	80
		0–4	316
		5+	52
		0–8	213
		9+	156
	ISS	0–15	248
		16+	121
		0–24	277
		25+	92
	Fatalities	Not fatal	318
		Fatal	51

Table 5.17 Multivariate logistic regression models for ISS9+, ISS16+, and ISS25+, age group 4+ (PCDS)

Variable	Symbol	Scaling factor	p-value	Adjusted odds ratio	95 % CI	
ISS9+ (AIC: 318; BIC: 334)						
Impact speed	$v_{c,PCDS}$	28.9	<0.001	26.197	12.942	53.028
Age	y_{ped}	22.2	<0.001	1.958	1.455	2.636
Hip height to forward hood height	$r_{4,PCDS}$	0.3	0.016	0.703	0.529	0.936
Constant	$\exp(\beta_0)$			0.030		
ISS16+ (AIC: 286; BIC: 302)						
Impact speed	$v_{c,PCDS}$	28.9	<0.001	21.256	10.975	41.169
Age	y_{ped}	22.2	<0.001	1.866	1.376	2.530
Hip height to transition point h.	$r_{3,PCDS}$	0.2	0.004	0.618	0.446	0.855
Constant	$\exp(\beta_0)$			0.017		
ISS25+ (AIC: 221; BIC: 241)						
Impact speed	$v_{c,PCDS}$	28.9	<0.001	29.250	13.644	62.705
Age	y_{ped}	22.2	<0.001	2.114	1.490	2.998
Hip height to transition point h.	$r_{3,PCDS}$	0.2	<0.001	0.298	0.161	0.551
Shoulder height to front hood h. (wrap)	$r_{6,PCDS}$	0.4	0.040	1.639	1.023	2.627
Constant	$\exp(\beta_0)$			0.005		

adjusted odds ratio, whereas shoulder height to the distance from the ground to the front hood opening ($r_{6,PCDS}$) is associated with a large (1.639 > 1) adjusted odds ratio. These two factors are correlated (Spearman coefficient 0.817, $p < 0.001$). Considering the log-odds ratios, which are -1.209 and 0.494 respectively, the effects thus almost cancel each other out, with a negative odds ratio of about 0.489 associated with body height. There is also a correlation between hip and shoulder height and each of the vehicle characteristics and body height as well as between the two vehicle profile characteristics. That means that an increase in the persons height relative to the height of the vehicle front is still beneficial.

The variance underestimate due to imputation was quantified using BMI as variable including body weight ($m_{ped,PCDS}$) in the fatality model for the PCDS data set. Ten instances with a total of 20.0 % imputed values each for BMI have been computed, compared to 10.0 % in the original data. The RMS deviation of the regression coefficient was 4.80 %. As for the GIDAS data before, this deviation is much smaller than the confidence intervals resulting from logistic regression.

Table 5.18 Multivariate logistic regression models for MAIS2+, MAIS3+, MAIS4+, and MAIS5+, age group 4+ (PCDS)

Variable	Symbol	Scaling factor	p-value	Adjusted odds ratio	95% CI	
MAIS2+ (AIC: 377; BIC: 389)						
Impact speed	$v_{c,PCDS}$	28.9	<0.001	12.605	6.840	23.229
Age	y_{ped}	22.2	<0.001	1.715	1.312	2.243
Constant	$\exp(\beta_0)$			0.149		
MAIS3+ (AIC: 327; BIC: 343)						
Impact speed	$v_{c,PCDS}$	28.9	<0.001	17.135	9.216	31.858
Forward hood height at centerline	$h_{3,PCDS}$	17.0	0.031	1.329	1.026	1.721
Age	y_{ped}	22.2	<0.001	1.669	1.256	2.220
Constant	$\exp(\beta_0)$			0.035		
MAIS4+ (AIC: 229; BIC: 245)						
Impact speed	$v_{c,PCDS}$	28.9	<0.001	17.180	8.993	32.820
Age	y_{ped}	22.2	<0.001	1.809	1.297	2.521
Hip height to transition point h.	$r_{3,PCDS}$	0.2	0.001	0.483	0.319	0.729
Constant	$\exp(\beta_0)$			0.008		
MAIS5+ (AIC: 196; BIC: 207)						
Impact speed	$v_{c,PCDS}$	28.9	<0.001	10.531	5.809	19.093
Transition point height at contact	$h_{4,PCDS}$	15.6	0.001	1.897	1.301	2.767
Constant	$\exp(\beta_0)$			0.007		

Table 5.19 Multivariate logistic regression models for fatalities, age group 4+ (PCDS)

Variable	Symbol	Scaling factor	p-value	Adjusted odds ratio	95% CI	
Fatalities (AIC: 182; BIC: 202)						
Body mass index	BMI	5.7	0.002	1.742	1.225	2.479
Impact speed	$v_{c,PCDS}$	28.9	<0.001	11.558	6.125	21.810
Age	y_{ped}	22.2	<0.001	2.163	1.439	3.250
Pedestrian hip height	$h_{h,ped,PCDS}$	11.1	0.002	0.461	0.281	0.758
Constant	$\exp(\beta_0)$			0.004		

Using Eq. 5.1 the resulting models can be written as follows:

$$p_{ISS9+,PCDS} = \left(1 + \exp\left(3.505 + 0.352 \cdot r_{4,PCDS} - 0.672 \cdot y_{ped} - 3.266 \cdot v_{c,PCDS}\right)\right)^{-1} \tag{5.24}$$

$$p_{ISS16+,PCDS} = \left(1 + \exp\left(4.048 + 0.482 \cdot r_{3,PCDS} \right.\right.$$
$$\left.\left. -0.624 \cdot y_{ped} - 3.057 \cdot v_{c,PCDS}\right)\right)^{-1} \tag{5.25}$$

Table 5.20 Predictive accuracy of the models given by Eqs. 5.24–5.27 (PCDS)

Model	Full-data model				Cross-validation			
	ROC AUC	95 % CI		k	ROC AUC	SD	SE	Optimism
ISS9+	0.880	0.846	0.914	3	0.871	0.060	0.020	0.009
ISS16+	0.890	0.858	0.923	3	0.880	0.078	0.026	0.011
ISS25+	0.901	0.868	0.935	4	0.900	0.052	0.017	0.002
MAIS2+	0.817	0.775	0.859	2	0.802	0.063	0.021	0.015
MAIS3+	0.870	0.834	0.905	3	0.850	0.048	0.016	0.020
MAIS4+	0.903	0.868	0.939	2	0.875	0.113	0.038	0.028
MAIS5+	0.898	0.856	0.940	2	0.877	0.069	0.023	0.021
Fatalities	0.913	0.876	0.950	4	0.898	0.058	0.019	0.014

$$p_{ISS25+,PCDS} = \left(1 + \exp\left(5.273 + 1.209 \cdot r_{3,PCDS} - 0.748 \cdot y_{ped}\right.\right.$$
$$\left.\left. -3.376 \cdot v_{c,PCDS} - 0.494 \cdot r_{6,PCDS}\right)\right)^{-1} \tag{5.26}$$
$$p_{Fatalities,PCDS} = \left(1 + \exp\left(5.47 - 0.771 \cdot y_{ped} + 0.774 \cdot h_{h,ped,PCDS}\right.\right.$$
$$\left.\left. -2.447 \cdot v_{c,PCDS} - 0.555 \cdot BMI\right)\right)^{-1} \tag{5.27}$$

The findings from the PCDS data are comparable to the GIDAS data set. Impact speed and pedestrian age are very important predictors in the multivariate models. Pedestrian and vehicle characteristics are significant as additional explanatory variables. As discussed in Sect. 5.3.1, an increased ratio of body height to vehicle front is beneficial. This effect is clearly present in the PCDS data and is masked for GIDAS due to a correlation between body height and impact speed.

The predictive performance of the models is quantified using ROC analysis. Table 5.20 summarizes the corresponding results. The predictive quality, in-sample and out-of-sample, is remarkably high (ROC AUC 0.802–0.900). The optimism is relatively small for all models.

Using the expected out-of-sample performance as measure, the ISS-based models tend to have higher mean AUC ROC (0.871–0.900) than the MAIS-based models (0.802–0.877). As in GIDAS, this is a clear indication that ISS has to be favored, considering the construction of injury probability models based on empirical in-depth accident data.

5.3.3 Multivariate Versus Univariate Analysis

The second research question refers to the expected advantage of multivariate compared to univariate modeling. To this end, univariate models for ISS and fatalities have been constructed using impact speed as single explanatory variable. The models for GIDAS are summarized in Table 5.21. The corresponding results for PCDS are summarized in Table 5.22 (note that impact speed was scaled differently for the PCDS data, see Sect. 5.2.3).

Table 5.21 Univariate results for ISS9+, ISS16+, ISS25+, and fatalities using impact speed as single predictor, age group 4+ (GIDAS)

Variable	Symbol	Scaling factor	p-value	Unadjusted odds ratio	95 % CI	
ISS9+ (speed only) (AIC: 754; BIC: 764)						
Impact speed	$v_{c,GIDAS}$	17.0	<0.001	3.622	2.889	4.541
Constant	$\exp(\beta_0)$			0.227		
ISS16+ (speed only) (AIC: 410; BIC: 420)						
Impact speed	$v_{c,GIDAS}$	17.0	<0.001	4.548	3.363	6.150
Constant	$\exp(\beta_0)$			0.056		
ISS25+ (speed only) (AIC: 316; BIC: 326)						
Impact speed	$v_{c,GIDAS}$	17.0	<0.001	3.107	2.335	4.134
Constant	$\exp(\beta_0)$			0.037		
Fatalities (speed only) (AIC: 271; BIC: 280)						
Impact speed	$v_{c,GIDAS}$	17.0	<0.001	3.995	2.939	5.429
Constant	$\exp(\beta_0)$			0.023		

Table 5.22 Univariate results for ISS9+, ISS16+, ISS25+, and fatalities using impact speed as single predictor, age group 4+ (PCDS)

Variable	Symbol	Scaling factor	p-value	Unadjusted odds ratio	95 % CI	
ISS9+ (speed only) (AIC: 340; BIC: 347)						
Impact speed	$v_{c,PCDS}$	28.9	<0.001	17.225	9.435	31.447
Constant	$\exp(\beta_0)$			0.045		
ISS16+ (speed only) (AIC: 308; BIC: 315)						
Impact speed	$v_{c,PCDS}$	28.9	<0.001	15.342	8.587	27.411
Constant	$\exp(\beta_0)$			0.025		
ISS25+ (speed only) (AIC: 255; BIC: 263)						
Impact speed	$v_{c,PCDS}$	28.9	<0.001	16.988	9.212	31.328
Constant	$\exp(\beta_0)$			0.012		
Fatalities (speed only) (AIC: 207; BIC: 215)						
Impact speed	$v_{c,PCDS}$	28.9	<0.001	8.098	4.789	13.693
Constant	$\exp(\beta_0)$			0.010		

The corresponding formulas using Eq. 5.1 are the following:

$$p_{ISS9+,speed,GIDAS} = \frac{1}{1 + \exp\left(1.484 - 1.287 \cdot v_{c,GIDAS}\right)} \tag{5.28}$$

$$p_{ISS16+,speed,GIDAS} = \frac{1}{1 + \exp\left(2.883 - 1.515 \cdot v_{c,GIDAS}\right)} \tag{5.29}$$

$$p_{ISS25+,speed,GIDAS} = \frac{1}{1 + \exp\left(3.288 - 1.134 \cdot v_{c,GIDAS}\right)} \tag{5.30}$$

Table 5.23 Predictive accuracy of the models containing only impact speed and comparison to multivariate results (GIDAS)

Model	Full-data model			Cross-validation					
	ROC AUC	95% CI		ROC AUC	SD	SE	Optimism	p value	t value
ISS9+	0.787	0.751	0.823	0.786	0.035	0.012	0.001	0.004	3.397
ISS16+	0.849	0.808	0.890	0.849	0.059	0.020	0.000	0.070	1.619
ISS25+	0.827	0.773	0.881	0.827	0.065	0.022	0.000	0.125	1.228
Fatalities	0.864	0.806	0.923	0.866	0.105	0.035	−0.002	0.020	2.395

$$P_{fatal,speed,GIDAS} = \frac{1}{1 + \exp\left(3.758 - 1.385 \cdot v_{c,GIDAS}\right)} \tag{5.31}$$

$$P_{ISS9+,speed,PCDS} = \frac{1}{1 + \exp\left(3.111 - 2.846 \cdot v_{c,PCDS}\right)} \tag{5.32}$$

$$P_{ISS16+,speed,PCDS} = \frac{1}{1 + \exp\left(3.674 - 2.731 \cdot v_{c,PCDS}\right)} \tag{5.33}$$

$$P_{ISS25+,speed,PCDS} = \frac{1}{1 + \exp\left(4.465 - 2.833 \cdot v_{c,PCDS}\right)} \tag{5.34}$$

$$P_{fatal,speed,PCDS} = \frac{1}{1 + \exp\left(4.625 - 2.092 \cdot v_{c,PCDS}\right)} \tag{5.35}$$

Tables 5.23 and 5.24 give the ROC AUC for the univariate models for GIDAS and PCDS. The in-sample and out-of-sample predictive accuracy of the models is high. Regarding the latter one, the models derived from PCDS tend to be more accurate. The optimism is very small (≤ 0.002). One-sided t-tests were used to evaluate the differences between cross-validated multivariate models (as given in Tables 5.15 and 5.20) and the corresponding univariate models.

The p-values in the last column refer to the hypothesis of improved ROC AUC in the cross-validated *multivariate* models. The hypothesis cannot be accepted for every model due to non-significant ($p < 0.05$) differences in the mean. In the GIDAS data set, the hypothesis can be accepted for ISS9+ ($p = 0.004$) and fatalities ($p = 0.020$). Regarding the non-significant models, there is a clear trend towards the multivariate models. As the standard deviation as well as the standard error of the mean are much greater than the optimism, it can be suspected that more data would be beneficial for proving a significant effect in every model. On the basis of a clear trend and the small (or significant) p-values, it is assumed that multivariate modeling indeed is beneficial compared to univariate modeling using impact speed of the vehicle alone.

There is one model in the literature which predicts mortality based on impact speed only [6]:

Table 5.24 Predictive accuracy of the models containing only impact speed and comparison to multivariate results (PCDS)

Model	Full-data model			Cross-validation					
	ROC AUC	95 % CI		ROC AUC	SD	SE	Optimism	p value	t value
ISS9+	0.855	0.816	0.894	0.853	0.049	0.016	0.002	0.084	1.496
ISS16+	0.870	0.832	0.909	0.871	0.084	0.028	−0.001	0.251	0.700
ISS25+	0.881	0.838	0.924	0.881	0.054	0.018	0.001	0.079	1.542
Fatalities	0.876	0.826	0.926	0.874	0.079	0.026	0.002	0.114	1.291

$$P_{fatal} = \frac{1}{1 + \exp\left(6.9 - 0.090 \cdot v_c\right)} \tag{5.36}$$

Again, the coefficients of the models are not directly comparable, as each model uses a different scaling for impact speed. Note, that the Rosen model used kph, whereas impact speed for the GIDAS model was scaled using standard deviation and mean and for PCDS using only the mean, see Sect. 5.2.3.

Figure 5.4 gives the mortality probability of the three models (Eqs. 5.31, 5.35, 5.36) as function of vehicle impact speed. The model constructed in this thesis based on GIDAS as well as the model by Rosen give rather similar estimates. As both models are constructed using the GIDAS data set and include the same parameters, this is a plausible and expected finding. The minor differences between the models could be attributable to, for example, differences in methodological aspects, such

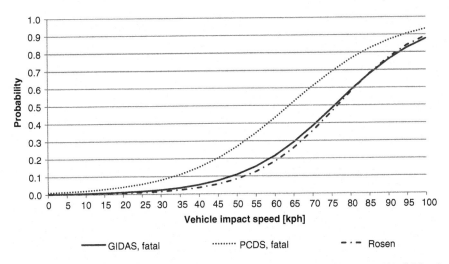

Fig. 5.4 Mortality estimates depending on vehicle impact speed, as given by Eqs. 5.31, 5.35 and 5.36

as imputation carried out here. The PCDS-based model shows higher probabilities
for given values of impact speed. This could be attributed to the PCDS data sample,
which represents an older situation from the 1990s (see Sect. 5.2.1). In addition to
that, differences in the safety features of the US vehicle population of that time, e.g.,
proportion of light trucks or structural characteristics, could explain higher injuries
at comparable levels of impact speed.

5.3.4 Investigation of Special Subgroups

The construction of injury probability models can include the whole data set, as presented above, or can be performed using subgroups of the data set. In view of biomechanical differences [19–21], a splitting of the population into subgroups depending
on age (e.g., 4–17, 18–64, 65+) could help detect possible injury risk factors specific
to particular age groups and improve the model quality.

Considering the limited number of cases available, a splitting into subgroups
reduces the statistical power and leads to the exclusion of some groups from analysis. Tables 5.25 and 5.26 give the frequencies of cases available for GIDAS and
PCDS. Groups with an asterisk (*) are considered insufficient for multivariate analysis (see Sect. 5.2.5) as they have less than 25 cases in at least one category. The low
numbers available, especially in categories containing more severe cases, highlight
the problem of limited case numbers in these data sets.

Table 5.27 gives the models for the GIDAS data set per age group. As the case
numbers are very low in some groups (especially for ISS16+, fatalities and the ado-

Table 5.25 Cases available in the GIDAS data set by outcome category and age group

Injury level	Age group					
	4–17		18–64		65+	
ISS9+	36	(272)	84	(314)	75	(96)
ISS16+	11	(297) *	39	(359)	35	(136)
ISS25+	4	(304) *	24	(374) *	23	(148) *
Fatalities	2	(317) *	22	(394) *	25	(155)

Table 5.26 Cases available in the PCDS data set by outcome category and age group

Injury level	Age group					
	4–17		18–64		65+	
ISS9+	28	(55)	100	(133)	28	(25)
ISS16+	19	(64) *	78	(155)	24	(29) *
ISS25+	14	(69) *	58	(175)	20	(33) *
Fatalities	6	(77) *	29	(204)	16	(37) *

Table 5.27 Multivariate logistic regression models by age groups (GIDAS)

Variable	Symbol	Scaling factor	p-value	Adj. OR	95 % CI	
ISS9+ (age 4-17) (AIC: 154; BIC: 162)						
Impact speed (sq.)	$v^2_{c,GIDAS}$	1473.5	<0.001	4.094	2.256	7.430
Constant	$\exp(\beta_0)$			0.110		
ISS9+ (age 18-64) (AIC: 277; BIC: 285)						
Impact speed	$v_{c,GIDAS}$	17.0	<0.001	3.663	2.556	5.251
Constant	$\exp(\beta_0)$			0.191		
ISS9+ (age 65+) (AIC: 164; BIC: 174)						
Pontoon-shaped	$type_{2,veh,GIDAS}$	–	0.033	4.399	1.125	17.195
Impact speed	$v_{c,GIDAS}$	17.0	<0.001	4.026	2.364	6.856
Constant	$\exp(\beta_0)$		0.015	0.201		
ISS16+ (age 18-64) (AIC: 146; BIC: 178)						
Age	y_{ped}	25.8	0.003	4.455	1.661	11.947
Impact speed (sq.)	$v^2_{c,GIDAS}$	1473.5	<0.001	4.345	2.618	7.212
Lower bumper ref. line—long.	$LBRL_{l,GIDAS}$	2.9	0.024	1.827	1.083	3.081
Upper bumper ref. line—vert.	$UBRL_{v,GIDAS}$	4.0	<0.001	2.886	1.630	5.112
Bonnet lead. e.—l.	$BLE_{l,GIDAS}$	3.0	0.003	3.644	1.539	8.630
Height to hood rear end—vert.	$r_{4,GIDAS}$	0.2	0.001	8.022	2.365	27.211
Angle up. bum. ref. l. to bonnet lead. e.	$\alpha_{1,GIDAS}$	7.0	0.034	2.510	1.073	5.875
Constant	$\exp(\beta_0)$			0.008		
ISS16+ (age 65+) (AIC: 111; BIC: 123)						
Impact speed (sq.)	$v^2_{c,GIDAS}$	1473.5	<0.001	3.921	2.027	7.586
Lower bumper reference line—vert.	$LBRL_{v,GIDAS}$	9.2	0.024	0.433	0.209	0.896
Bonnet lead. e.—l.	$BLE_{l,GIDAS}$	3.0	0.003	2.667	1.395	5.099
Constant	$\exp(\beta_0)$			0.121		
Fatalities (age 65+) (AIC: 83; BIC: 99)						
Body weight	$m_{ped,GIDAS}$	21.3	0.004	8.061	1.959	33.175
Age	y_{ped}	25.8	0.007	32.719	2.596	412.406
Kinetic energy	E_{kin}	1008.5	<0.001	3.564	1.801	7.052
Bonnet lead. e.—l.	$BLE_{l,GIDAS}$	3.0	0.008	2.868	1.313	6.262
Constant	$\exp(\beta_0)$			0.000		

lescent group with ISS9+). Statistical effects such as multicollinearity and selection effects strongly influence the results and make a sound interpretation of the explanatory factors difficult. One manifestation can be that the relative ordering of predictor importance is no longer reliable.

The selection of subgroups can produce correlations between factors which are not correlated in the whole data set. The effect is obvious in the model for fatalities (age 65+). Pedestrian age as well as weight are of overriding importance compared to impact speed (included in kinetic energy). The correlations between the variables do not allow for an interpretation of effects concerning cause-effect relations. Kinetic energy in this model is significantly correlated with age and weight. Both explanatory variables are also correlated with linear impact speed (which is not included in the multivariate model) and thus are suspected to be confounders within this special subgroup.

As a result, the further splitting of the whole data set seems to reduce power and produce selection effects, which distort the results and make a sound interpretation nearly impossible. The statistical problems become also evident in very large confidence intervals. The method elaborated in the preceding sections can therefore not be recommended for small empirical data sets.

The corresponding results for PCDS are given in Table 5.28. As many of the groups presented have low case numbers, similar effects as presented above for the GIDAS data become evident for the PCDS data set. The large confidence intervals are an indication for the instability of the models. A further interpretation of the results does not seem feasible and is therefore not given.

Although the splitting of the population with respect to pedestrian age is grounded on biomechanical considerations, a splitting of the population in order to investigate special subgroups does not seem to make sense using the data sets available: The case numbers are too low to construct about half of the models in question. Case numbers are critically low (as explained above) for most of the other models shown. The instability of the models combined with selection effects does not allow a clear interpretation of the results and a comparison with the multivariate results for the whole data set regarding predictive accuracy (as presented in Sect. 5.3.2). Nevertheless, pedestrian age is included in the multivariate analysis of Sect. 5.3.2 and shows up as significant explanatory variable in every ISS and fatality model in both GIDAS and PCDS. Thus, the effects of pedestrian age are considered and quantified in the models there.

5.4 Plausibility Check and Indications for Implementation

5.4.1 Probability Models for ISS and Fatalities

In order to check the plausibility of the GIDAS-based models for ISS (univariate and multivariate) as given by Eqs. 5.17–5.19, p. 120, and 5.28–5.30, p. 127, the conditional probability simulation described in Sect. 5.2.6.1 is used. The necessary parameters required by the models are v_c, y_{ped}, m_{ped}, $LBRL_v$, BLE_v, and r_1.

Table 5.28 Multivariate logistic regression models by age groups (PCDS)

Variable	Symbol	Scaling factor	p-value	Adj. OR	95 % CI	
ISS9+ (age 4–17) (AIC: 72; BIC: 77)						
Impact speed	$v_{c,PCDS}$	28.9	<0.001	44.573	7.604	261.281
Constant	$exp(\beta_0)$			0.009		
ISS9+ (age 18–64) (AIC: 205; BIC: 218)						
Impact speed	$v_{c,PCDS}$	28.9	<0.001	22.002	9.514	50.879
Forward hood height at centerline	$h_{3,PCDS}$	17.0	0.019	1.426	1.061	1.917
Age	y_{ped}	22.2	0.007	2.356	1.257	4.416
Constant	$exp(\beta_0)$			0.035		
ISS9+ (age 65+) (AIC: 49; BIC: 53)						
Impact speed	$v_{c,PCDS}$	28.9	<0.001	29.566	4.483	194.989
Constant	$exp(\beta_0)$			0.072		
ISS16+ (age 18–64) (AIC: 160; BIC: 171)						
Impact speed	$v_{c,PCDS}$	28.9	<0.001	38.631	14.584	102.329
Hip height to transition point height	$r_{3,PCDS}$	0.2	0.001	0.445	0.279	0.710
Constant	$exp(\beta_0)$			0.011		
ISS25+ (age 18–64)(AIC: 114; BIC: 131)						
Vehicle curb weight	$m_{veh,PCDS}$	1415.1	<0.001	3.151	1.680	5.909
Impact speed	$v_{c,PCDS}$	28.9	<0.001	88.215	22.692	342.934
Forward hood height at centerline	$h_{3,PCDS}$	17.0	0.031	1.864	1.058	3.286
Angle of front bumper	$\alpha_{1,PCDS}$	15.4	<0.001	0.292	0.158	0.539
Constant	$exp(\beta_0)$			0.001		
Fatalities (age 18–64) (AIC: 160; BIC: 171)						
Impact speed	$v_{c,PCDS}$	28.9	<0.001	37.118	4.953	278.154
Age	y_{ped}	22.2	0.043	36.807	1.120	1209.161
Constant	$exp(\beta_0)$			0.000		

Using the conditional probability simulation for the GIDAS-based ISS models, it becomes obvious that constraint 2 (see Eq. 5.8, p. 101) can actually be violated. Considering the models, this means that

- $p_{ISS16+,GIDAS} \leq p_{ISS9+,GIDAS}$ is violated about in 2.0 % of the cases,
- $p_{ISS25+,GIDAS} \leq p_{ISS9+,GIDAS}$ is violated in 0.0 % of the cases,
- $p_{ISS25+,GIDAS} \leq p_{ISS16+,GIDAS}$ is violated about in 31.1 % of the cases,
- $p_{ISS16+,speed,GIDAS} \leq p_{ISS9+,speed,GIDAS}$ is violated in 0.0 % of the cases,
- $p_{ISS25+,speed,GIDAS} \leq p_{ISS9+,speed,GIDAS}$ is violated in 0.0 % of the cases, and
- $p_{ISS25+,speed,GIDAS} \leq p_{ISS16+,speed,GIDAS}$ is violated about in 14.4 % of the cases.

Table 5.29 Frequencies of target variables (GIDAS)

Injury Level		Cases
	9–15	110
	16+	85
ISS	16–24	34
	25+	51
	0–8	682
	9–15	110

Table 5.30 Univariate results for ISS16+|9+, and ISS25+|16+, using impact speed as single predictor, age group 4+ (GIDAS), Option a

Variable	Symbol	Scaling factor	p-value	Unadjusted odds ratio	95% CI	
ISS16+|9+ (speed only) (AIC: 237; BIC: 244)						
Impact speed	$v_{c,GIDAS}$	17.0	<0.001	2.448	1.708	3.507
Constant	$\exp(\beta_0)$			0.391		
ISS25+|16+ (speed only)						
Constant	$\exp(\beta_0)$			0.067		

As, for example, ISS9+ and ISS25+ are rather far apart, a violation of constraint 2 does not arise within the parameter ranges considered, neither multi- nor univariate. ISS16+ and ISS25+ are much closer which leads to the described violation.

This section includes the results for the models constructed with the approach of conditional probabilities as explained above. In order to demonstrate the method, Options a and c (see p.106) are computed each for the univariate and multivariate GIDAS models. Option c leaves the original ISS16+ model with its wealth of explanatory factors (especially vehicle characteristics). Option a gives insight regarding a generic approach. Option b is not given, as it does not bring any new insights into the application of the methodology.

Table 5.29 gives the numbers available for analysis in the GIDAS data set. Due to the selection necessary for constructing the new models, the numbers are smaller than the those in Table 5.11, p. 117.

The corresponding results for Option a are given in Table 5.30 for the univariate case and in Table 5.31 for the multivariate one.

Using Eq. 5.1 the resulting models can be written as:

$$p_{ISS16+|9+,speed,GIDAS} = \frac{1}{1 + \exp\left(0.939 - 0.895 \cdot v_{c,GIDAS}\right)} \tag{5.37}$$

$$p_{ISS25+|16+,speed,GIDAS} = \frac{1}{1 + \exp\left(-0.405\right)} \tag{5.38}$$

$$p_{ISS16+|9+,GIDAS} = \left(1 + \exp\left(0.637 - 0.997 \cdot v_{c,GIDAS} - 0.983 \cdot BMI - \right.\right.$$
$$\left.\left. -0.667 \cdot LBRL_{v,GIDAS} \, 1.435 \cdot W_{1,GIDAS}\right)\right)^{-1} \tag{5.39}$$

Table 5.31 Multivariate results for ISS16+|9+, and ISS25+|16+, age group 4+ (GIDAS), Option a

Variable	Symbol	Scaling factor	p-value	Adj. OR	95 % CI	
ISS16+|9+ (AIC: 176; BIC: 186)						
Impact speed	$v_{c,GIDAS}$	17.0	<0.001	2.710	1.778	4.130
Body mass index	BMI	5.2	<0.001	2.673	1.582	4.517
Lower bumper	$LBRL_{v,GIDAS}$	9.2	0.005	0.529	0.339	0.826
reference line—vert.						
Bonnet lead. edge (wrap)	$W_{1,GIDAS}$	6.6	0.023	1.949	1.098	3.457
Constant	$\exp(\beta_0)$			0.238		
ISS25+|16+						
Constant	$\exp(\beta_0)$			0.051		

$$p_{ISS25+|16+,GIDAS} = \frac{1}{1 + \exp(-0.495)} \tag{5.40}$$

The models described above are used to compute composite speed models $p_{ISS16+,speed,GIDAS,a}$ using Eqs. 5.11 together with 5.28 and $p_{ISS25+,speed,GIDAS,a}$ using 5.12 together with $p_{ISS16+,speed,GIDAS,a}$. The multivariate models follow the same scheme (of course using Eq. 5.17 instead of Eq. 5.28). The composite models are tested using the conditional probability simulation described. It is easily verified that constraint 2 is fulfilled 100.0 %.

The predictive performance of the models is again quantified using ROC AUC. The underlying question is, whether the proposed new method of constructing the model results in a loss of predictive accuracy. To this end, the new models of Tables 5.30 and 5.31 are compared with the ones of Sect. 5.3 once within their subset of data and then as complete models on the whole data set.

Table 5.32 gives the corresponding results for the *univariate* models for Option *a*. The predictive accuracy within the subset of the data is identical for the models constructed on the subset compared to the ones constructed on the full data set. For ISS16+|9+, the ROC AUC is identical. For ISS25+|16+ both models do not have any predictive power. The reason for this can be low case numbers as well as a close proximity of the outcome categories themselves. A good indication for this is also the ISS25+|16+ model itself. Within the ISS16+ subset, even impact speed does not have sufficient explanatory power for ISS25+, so the model consists only of a constant. The constant indicates the probability for an ISS16+ injury being an ISS25+ injury being 62.1 % (Eq. 5.40), as no remaining explanatory factors can be identified. Comparing the composite model and the original model on the full data set results in the same predictive power for the models using only impact speed as predictor.

Table 5.33 gives the corresponding results for the *multivariate* models for Option *a*. The multivariate models do show minor differences regarding predictive power. The models constructed on a subset of the data have less predictive power (again, for

Table 5.32 Predictive accuracy of the models using impact speed as single predictor for Option a (GIDAS)

Model	Data set	ROC AUC	95 % CI		k	Difference
ISS16+|9+ (speed only)	ISS9+	0.718	0.646	0.790	1	0.000
ISS16+ (speed only)	ISS9+	0.718	0.646	0.790	1	
ISS16+$_a$ (speed only)	Full data set	0.849	0.808	0.890	1*	0.000
ISS16+ (speed only)	Full data set	0.849	0.808	0.890	1	
ISS25+|16+ (speed only)	ISS16+	0.500	0.374	0.626	0	0.022
ISS25+ (speed only)	ISS16+	0.478	0.354	0.603	1	
ISS25+$_a$ (speed only)	Full data set	0.827	0.773	0.881	1*	0.000
ISS25+ (speed only)	Full data set	0.827	0.773	0.881	1	

The number of included parameters is given by k. An asterisk (*) indicates that k is the sum of different parameters included in the models used. The difference refers to both ROC AUC values

Table 5.33 Predictive accuracy of the multivariate models for Option a (GIDAS)

Model	Data set	ROC AUC	95 % CI		k	Difference
ISS16+|9+	ISS9+	0.801	0.734	0.869	4	−0.017
ISS16+	ISS9+	0.819	0.756	0.882	6	
ISS16+$_a$	Full data set	0.912	0.881	0.943	5*	−0.009
ISS16+	Full data set	0.921	0.891	0.950	6	
ISS25+|16+	ISS16+	0.500	0.374	0.626	0	−0.079
ISS25+	ISS16+	0.579	0.456	0.702	2	
ISS25+$_a$	Full data set	0.896	0.854	0.938	5*	0.008
ISS25+	Full data set	0.888	0.843	0.932	2	

The number of included parameters is given by k. An asterisk (*) indicates that k is the sum of different parameters included in the models used. The difference refers to both ROC AUC values

ISS25+, both models do not have any predictive power within the ISS16+ subset). The ISS16+ composite model has a marginally lower ROC AUC compared to the full data model. For ISS25+, it is vice versa.

The multivariate results for Option c are given in the following. As explained above, the results for Option b are not explicitly given. The ISS16+ model (Eq. 5.18) includes six explanatory factors and therefore is kept in Option c. Table 5.34 includes the corresponding models (note that the ISS25+|16+ model is the same as given in Table 5.31).

Using Eq. 5.1 the resulting model can be written as:

$$p_{ISS9+|15-,GIDAS} = \frac{1}{1 + \exp\left(1.887 - 0.597 \cdot y_{ped} - 1.006 \cdot v_{c,GIDAS}\right)} \quad (5.41)$$

The model is used to compute the composite models $p_{ISS9+,GIDAS,c}$, using Eq. 5.15 together with 5.18. The composite model $p_{ISS25+,GIDAS,c}$ is computed using 5.16

Table 5.34 Multivariate results for ISS9+|15−, age group 4+ (GIDAS)

Variable	Symbol	Scaling factor	p-value	Adjusted odds ratio	95 % CI	
ISS9+|15− (AIC: 496; BIC: 510)						
Age	y_{ped}	25.8	<0.001	1.816	1.444	2.284
Impact speed	$v_{c,GIDAS}$	17.0	<0.001	2.735	2.060	3.631
Constant	$\exp(\beta_0)$			0.152		

Table 5.35 Predictive accuracy of the multivariate models for Option c (GIDAS)

Model	Data set	ROC AUC	95 % CI		k	Difference
ISS9+|15−	ISS15−	0.761	0.712	0.809	2	0.000
ISS9+	ISS15−	0.760	0.712	0.808	2	
ISS9+$_c$	Full data set	0.824	0.788	0.860	6*	−0.001
ISS9+	Full data set	0.825	0.789	0.861	2	
ISS25+|16+	ISS16+	0.500	0.374	0.626	0	−0.079
ISS25+	ISS16+	0.579	0.456	0.702	2	
ISS25+$_c$	Full data set	0.906	0.865	0.946	6*	0.018
ISS25+	Full data set	0.888	0.843	0.932	2	

The number of included parameters is given by k. An asterisk (*) indicates that k is the sum of different parameters included in the models used. The difference refers to both ROC AUC values

together with 5.19 and 5.40. The composite models are tested again using the conditional probability simulation described. Constraint 2 is fulfilled 100.0 %.

Table 5.35 gives the results for explanatory power by ROC AUC (the values for ISS25+|16+ are identical to the ones in Table 5.33). As for Option a, also Option c leads to insignificant differences in the ROC AUC. Both options thus produce models with high in-sample predictive accuracy and without any implausibilities, independent of the actual values of parameters inserted.

5.4.2 Implications and Conclusion on Plausibility

In order to ensure plausibility of injury probability models, two important constraints are defined (Sect. 5.2.6): First, the probability is defined as zero if $v_c = 0$ kph. Second, the probabilities for an outcome variable which itself is a subset of another outcome variable (e.g., ISS25+ and ISS16+; $ISS25+ \subseteq ISS16+$) must not be greater than the corresponding probability for the other outcome variable (e.g., $p_{ISS25+} \leq p_{ISS16+}$).

A conditional probability simulation using all relevant parameters was used to test constraint 2 (high case numbers and Monte-Carlo techniques ensured a testing of a very large combination of parameters). The models of Sect. 5.3 were found to violate that constraint. The solution chosen uses conditional probability identity and therefore requires the construction of new models. Different approaches concerning starting points are presented and two options are explicitly discussed as examples. The method delivers models with comparably high predictive power compared to the models in Sect. 5.3. The drawback of the method is a reduction in case numbers for the construction of the new models. However, in the examples displayed, this does not lead to a reduced ROC AUC for the models, but becomes evident in newly constructed ISS25+|16+ models, which do only consist of a constant and do not include any further variables.

In general there are several implications for the practical use of injury probability models and consequently of the method as presented. The correct implementation of the models and the meaning of constraint 2 depend on the research question of the study:

- Constraint 2 is meaningless, if only one outcome category is to be assessed (e.g., fatalities).
- Constraint 2 has to be considered, if more than one outcome category is to be assessed (e.g., ISS0-8, ISS9-14, and ISS16+). In case of more than two outcome categories, a model has to be selected as starting point for the development of the other models with respect to constraint 2. This can in principle be any of the models in question, which subsequently stays unchanged.

In the example above, Option c was favored as starting model (ISS16+) due to the high explanatory value concerning number and kind of factors instead of the more generic Options a and b. Besides, other decision criteria for selecting the starting model could be ROC AUC values or explanatory factors included (or not) in the *newly* constructed models.

5.5 Conclusion

Methodology for the construction of different self-consistent probability models for injury level as well as fatalities has been developed and applied to pedestrians in frontal vehicle crashes in this chapter. In-depth accident data from Germany (GIDAS) and the US (PCDS) were used. The number of cases available in each data set is low; GIDAS provides roughly three times the number of cases than PCDS. Whereas PCDS is a project which was finished in the 1990s, GIDAS has been collecting and reconstructing cases for about the last ten years and it is an ongoing project.

The different procedures regarding data preparation and analysis leading to the probability models are described in detail. In order to assure a maximum of possible statistical power and integrity of the results, the data were checked for consistency and missing data were imputed. The interpretation of effect sizes is simplified by standardization of the continuous variables used. Non-continuous variables were recoded into binaries to be compatible with logistic regression. Predictive quality of the models was assessed using ROC analysis. As there are no external data available for validation of the models, 10-fold cross-validation was used to evaluate the expected out-of-sample predictive quality and check for possible over-fitting due to multivariate modeling.

The difficulties while working with empirical (observational) data are evident for the data used. Issues of potential confounding factors, multicollinearity, and selection effects are addressed. This approach is clearly limited by low case numbers.

Three hypotheses were investigated while constructing the models. The first is based on medical literature and refers to the advantage of ISS as overall injury metric over MAIS. As expected, there is a clear trend that ISS-based models are more accurate than MAIS-based models. The second hypothesis investigated the difference between univariate modeling (based on vehicle impact speed) and multivariate modeling. As a result, the multivariate models show a clear trend to be more accurate (for some of them the advantage is significant). The third hypothesis is again based on medical literature and refers to individual modeling for specific age groups (for the pedestrian). The last hypothesis cannot be tested, as the number of cases is too low for this kind of statistical analysis and thus produces distorting effects as well as a severe loss in statistical power. More data would be needed to answer that research question. Again, the importance of data preparation in terms of imputation becomes obvious.

Previous findings documented in the literature are confirmed by the results. Impact speed is by far the most important predictor for injury severity and mortality (in both data sets). Pedestrian age is also included in every model. Vehicle characteristics (i.e., geometric quantities of the front end) as well as pedestrian physiology (e.g., BMI) are also significant in multivariate analysis.

As explained above, self-consistent models for several outcome categories require special care in order to satisfy the constraints imposed by the laws of probability. Failure to take these constraints into account could lead to contradictory results, particularly in multiple regression models when extreme values of explanatory variables are considered. To solve this problem, a method using conditional probability identities was developed and applied; this method seems to be novel in the context of risk analysis for vehicle safety. The results are models that deliver self-consistent results for every possible combination of explanatory variables as well as number of outcome categories.

Considering the high priority of pedestrian protection among European agencies and in the international safety community, as well as the resources devoted to theoretical and political discussion of the subject, it seems surprising (to say the least) that empirical data resources are so scarce. Thus, larger and more recent data sets with representative sampling from multiple countries are highly important for better characterization of factors influencing pedestrian risks and ultimately for optimization of active and integrated pedestrian protection systems. Injury severity is a metric which is capable of comparing both passive and active safety approaches on the same scale. The methodology explained and the resulting models derived in this section provide the basis for an objective and quantitative evaluation of preventive pedestrian protection measures.

References

1. German In-Depth Accident Study: Unfalldatenbank 07.1999-12.2008. Dresden, Hannover, 31.12.2008.
2. UMTRI (2005). *1994–1998 NASS pedestrian crash data study (PCDS) codebook. Version 03Mar01*. UMTRI Transportation Data Center.
3. Chauvel, C., Page, Y., Fildes, B., & Lahausse, J. (2013). Automatic emergency braking for pedestrians effective target population and expected safety benefits. In *23rd international technical conference on the enhanced safety of vehicles* (*ESV 2013*) No. 13–0008.
4. Hannawald, L., & Kauer, F. (2004). *Equal effectiveness study on pedestrian protection*. Dresden: Technische Universität Dresden.
5. Ressle, A., Schramm, S., & Kölzow, T. (2010). Generierung von Verletzungsrisikofunktionen für Fußgängerkollisionen. In *Crash Tech 2010—Fahrzeugsicherheit 2020*.
6. Rosen, E., & Sander, U. (2009). Pedestrian fatality risk as a function of car impact speed. *Accident Analysis and Prevention, 41*, 536–542.
7. Tefft, B. (2011). *Impact speed and a pedestrian's risk of severe injury or death*. Report, AAA Foundation for Traffic Safety.
8. AAAM (1990). *The abbreviated injury scale (1990 revision)*. Association for the Advancement of Automotive Medicine.
9. States, J. D. (1969). Abbreviated and the comprehensive research injury scales. In *Proceedings of thirteenth Stapp Car Crash conference, December 2–4, 1969* (Vol. 13, pp. 282–294). New York: Society of Automotive Engineers.
10. States, J. D., & Huelke, D. F. (1980). *The abbreviated injury scale. 1980 revision*. Des Plaines: American Association for Automotive Medicine (AAAM).
11. The Abbreviated Injury Scale, Update (2008). Des Plaines, IL: American Association for Automotive Medicine (AAAM).
12. Codebook GIDAS2009. GIDAS, 2009.
13. Otte, D., Haasper, C., & Krettek, C. (2006 October). Die neue abbreviated injury scale (AIS) 2005—Nutzen einer standardisierten Klassifikation der Verletzungsschwere. *Verkehrsunfall und Fahrzeugtechnik (VKU)* (pp. 261–268).
14. Kramer, F. (Ed.). (2006). *Passive Sicherheit von Kraftfahrzeugen*. Wiesbaden: Friedr. Vieweg & Sohn Verlag.
15. Baker, S. P., & O'Neill, B. (1976). The injury severity score: An update. *The Journal of Trauma, 16*(11), 882–885.
16. Baker, S. P., O'Neill, B., Haddon, W. J., & Long, W. B. (1974). The injury severity score: A method for describing patients with multiple injuries and evaluating emergency care. *The Journal of Trauma, 14*(3), 187–196.
17. Nogueira, L. S., Domingues, C. A., Campos, M. A., & Ten Sousa, R. M. C. (2008). Years of new injury severity score (NISS): Is it a possible change? *Rev Latino-am Enfermagem, 16*(2), 314–319.
18. Stevenson, M., Segui-Gomez, M., Lescohier, I., Di Scala, C., & McDonald-Smith, G. (2001). An overview of the injury severity score and the new injury severity score. *Injury Prevention, 7*, 10–13.
19. Henary, B. Y., Crandall, J., Bhalla, K., Mock, C. N., & Roudsari, B. S. (2003). Child and adult pedestrian impact: The influence of vehicle type on injury severity. In *47th annual conference of the association for the advancement of automotive medicine*.
20. Henary, B. Y., Ivarsson, B. J., & Crandall, J. R. (2006). The influence of age on the morbidity and mortality of pedestrian victims. *Traffic Injury Prevention, 7*, 182–190.
21. Prange, M., Heller, M., Watson, H., Iyer, M., Ivarsson, B. J., & Fisher, J. (2010). Age effects on injury patterns in pedestrian crashes. *SAE International Journal of Passenger Cars—Mechanical Systems, 3*(1), 789–820.
22. GIDAS German In-Depth Accident Study (2010, December 22). http://www.gidas.org/files/GIDAS_eng.pdf.

23. Ebner, A., Samaha, R. R., Scullion, P., & Helmer, T. (2010). Identifying and analyzing reference scenarios for the development and evaluation of preventive pedestrian safety systems. In *Proceedings of the 17th ITS world congress*.
24. Ebner, A., Samaha, R. R., Scullion, P., & Helmer, T. (2010). Methodology for the development and evaluation of active safety systems using reference scenarios: Application to preventive pedestrian safety. In *Proceedings of the international research council on biomechanics of injury (IRCOBI)* (pp. 155–168).
25. Isenberg, R. A., Chidester, A. B., & Mavros, S. (1998). Update on the pedestrian crash data study. In *16th international technical conference on the enhanced safety of vehicles (ESV)*.
26. Jarrett, K. J., & Saul, R. A. (1998). Pedestrian injury-analysis of the PCDS field collision data. In *16th international technical conference on the enhanced safety of vehicles (ESV 1998)*.
27. Hakkert, A. S., Gitelman, V., & Vis, M. A. (2007). *Road safety performance indicators: Theory*. Deliverable D3.6, EU FP6 project SafetyNet.
28. Jaccard, J. (2001). *Interaction effects in logistic regression* (pp. 07–135). Thousand Oaks, CA: Sage University Papers Series on Quantitative Applications in the Social Sciences.
29. Statistisches, Bundesamt. (2010). *Mikrozensus—Fragen zur Gesundheit—Körpermaße der Bevölkerung—2009*. Wiesbaden: Statistisches Bundesamt.
30. Stolzenberg, H., Kahl, H., & Bergmann, K. (2007). Körpermaße bei Kindern und Jugendlichen in Deutschland. *Bundesgesundheitsblatt—Gesundheitsforschung—Gesundheitsschutz, 50*, 659–669. doi:10.1007/s00103-007-0227-5.
31. Liers, H., & Hannawald, L. (2008). *Klassifizierung von Fahrzeugfrontkonturen in Fußgängerfrontalunfällen*. Verkehrsunfallforschung an der TU Dresden GmbH: Abschlussbericht.
32. Ogden, C. L., Fryar, C. D., Carroll, M. D., & Flegal, K. M. (2004). Mean body weight, height, and body mass index, United States 1960–2002. Advance data from vital and health statistics 347. Hyattsville, Maryland: National Center for Health Statistics.
33. Clauß, G., Finze, F. -R., & Partzsch, L. (2011). *Grundlagen der statistik* (6th ed.). Frankfurt: Verlag Harri Deutsch.
34. Salkind, N. (2011). *Statistics for people who (think they) hate statistics* (4th ed.). Munich: SAGE Publications Inc.
35. Bortz, J., & Schuster, C. (2010). *Statistik für Human- und Sozialwissenschaftler* (7th ed.). Berlin: Springer.
36. Degen, H., & Lorscheid, P. (2012). *Statistik-Lehrbuch* (4th ed.). Munich: Oldenbourg Verlag München.
37. Kleinbaum, D., & Klein, M. (2010). *Logistic regression. A self learning text. Statistics for biology and health*. Berlin: Springer.
38. Tabachnick, B., & Fidell, L. (2013). *Using multivariate statistics* (6th ed.). London: Pearson.
39. Pampel, F. (2000). *Logistic regression: A primer* (pp. 07–132). Thousand Oaks, CA: Sage University Papers Series on Quantitative Applications in the Social Sciences.
40. Menard, S. (2002). *Applied logistic regression analysis* (2nd ed.). Thousand Oaks: SAGE Publications Inc.
41. Backhaus, K., Erichson, B., Plinke, W., & Weiber, R. (2008). *Multivariate analysemethoden* (12th ed.). Berlin: Springer.
42. Schwarz, G. (1978). Estimating the dimension of a model. *The Annals of Statistics, 6*(2), 461–464.
43. Findley, D. (1991). Counterexamples to parsimony and BIC. *Annals of the Institute of Statistical Mathematics, 43*(3), 505–514.
44. Kohavi, R. A. (1995). Study of cross-validation and bootstrap for accuracy estimation and model selection. In *International joint conference on artificial intelligence* (pp. 1137–1143).
45. Schramm, S. (2011). *Methode zur Berechnung der Feldeffektivität integraler Fußgängerschutzsysteme*. Dissertation, Technische Universität München.
46. Dahdah, S. (2008). Modeling an infrastructure safety rating for vulnerable road users in developing countries. Dissertation, The George Washington University.
47. Kerrigan, J. R., Rudd, R., Subit, D., Untaroiu, C. D., & Crandall, J. R. (2008). *Pedestrian lower extremity response and injury: A small sedan vs. a large SUV*. SAE Technical Paper, 2008-01-1245.

48. Lefler, D. E., & Gabler, H. C. (2001). *The emerging threat of light truck impacts with pedestrians*. SAE Technical Paper, 2001-06-0082.
49. Longhitano, D., Henary, B., Bhalla, K., Ivarsson, J., & Crandall, J. (2005). *Influence of vehicle body type on pedestrian injury distribution*. SAE Technical Paper, 2005-01-1876.
50. Roudsari, B. S., Mock, C. N., & Kaufman, R. (2005). An evaluation of the association between vehicle type and the source and severity of pedestrian injuries. *Traffic Injury Prevention, 6,* 185–192.
51. Simms, C. K., & Wood, D. P. (2006). Pedestrian risk from cars and sport utility vehicles—a comparative analytical study. In *Proceedings of the Institution of Mechanical Engineers* (Vol. 220, pp. 1085–1100).
52. Fitzharris, M., & Fildes, B. (2007). *Analysis of the potential crash reduction benefits of electronic brake assist, early warning systems, and the combined effects for pedestrians*. Melbourne, Australia: Monash University Accident Research Centre for the Automotive Collaborative Research Consortium.

Chapter 6
Integrated Evaluation of Preventive Pedestrian Protection

6.1 Design of Virtual Simulation Experiments: System Versus Reference

The proposed evaluation process for systems of integral safety, as described in Chap. 3, is based on virtual experiments comparing a system to a reference, which makes use of a variety of different data sources, modeling techniques as well as meaningful metrics. Although the details of the simulation itself are not part of this thesis (see Sect. 3.4), its results are described, discussed, and used to illustrate the practical application of the injury probability models developed in Chap. 5 and to highlight the overall methodology and process of integral safety evaluation.

The virtual experiments are designed to distinguish typical system effects corresponding to a few percent reduction of accidents. Thus, the number of accident events required for this level of precision is typically about 1,600 or more. This number would correspond to a standard deviation of ± 40 or 2.5 %, so that 5 % effectiveness changes could typically be seen. Higher precision is attainable with more events.

In the scenario of hazardous pedestrian crossing situations, about 0.2 % of the crossings (SD 0.004 %) result in a collision in the baseline. Hence, about one million crossings are usually simulated to resolve 5 % effects.

The version of the simulation used has the following distribution of key parameters. Baseline are 18 million crossings with corresponding accidents. Figure 6.1 gives the distribution of impact speed of the vehicles in a collision as well as the cumulative distribution of the corresponding fraction of accidents in GIDAS [1]. The GIDAS sample used for this comparison is described in Sect. 5.2.1 (p. 93). The speed distribution is plausible for the urban setting of the traffic situation. The maximum speed in the simulation is limited to 80 kph. The accidents in GIDAS have a trend towards lower speeds.

Pedestrian age and body height are further examples of important parameters (Figs. 6.2 and 6.3). The pedestrians in the GIDAS sample are younger and include also ages above 80 (80 is the maximum age for pedestrians in the simulation). Due to

© Springer International Publishing Switzerland 2015
T. Helmer, *Development of a Methodology for the Evaluation of Active Safety using the Example of Preventive Pedestrian Protection*, Springer Theses,
DOI 10.1007/978-3-319-12889-4_6

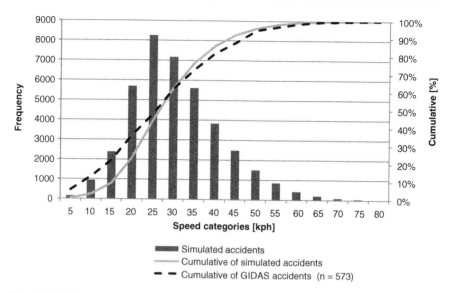

Fig. 6.1 Vehicle impact speeds in the baseline accidents and corresponding values from GIDAS

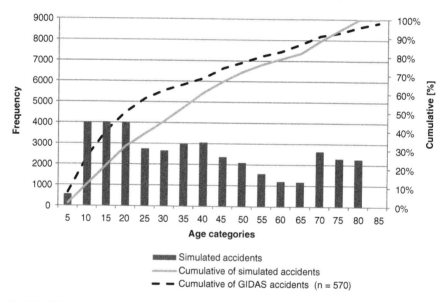

Fig. 6.2 Pedestrian age in the baseline accidents and corresponding values from GIDAS

the correlation of age and body height, smaller body heights are also more strongly represented in GIDAS than in the simulation.

Figure 6.4 gives the resulting injury probabilities for the models described in Sect. 5.2.6 (p. 101). Results include models depending only on impact speed of

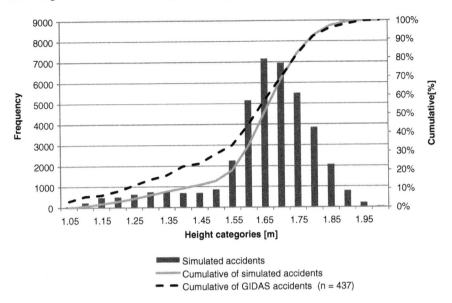

Fig. 6.3 Pedestrian body height in the baseline accidents and corresponding values from GIDAS

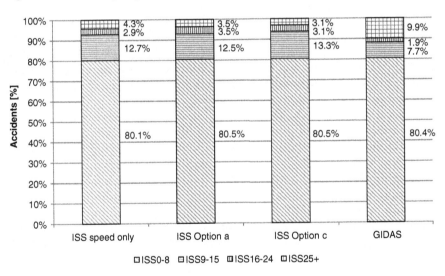

Fig. 6.4 Injury outcomes using different types of injury probability models from Sect. 5.2.6 and corresponding values from GIDAS

the vehicle (using Eqs. 5.28, 5.37, and 5.38), multivariate models using Option *a* (Eqs. 5.17, 5.39, and 5.40), Option *c* (Eqs. 5.41, 5.18, and 5.40), and the corresponding values from GIDAS. It is evident that each set of models lead to a different distribution of injury probabilities.

The exposure model in the simulation is based on US data. As the results in this section are intended as explanation of the methodology, the differences between the accident fraction of the simulation and the distributions derived from the corresponding accidents in GIDAS probably reflect differences in exposure relations between Germany and the US and do not affect the methodical considerations that follow. For a relative comparison between different system configurations (or vehicle configurations), this does not have any effect. Again, as also a functional demonstrator is used, the absolute values of the results given here are not intended for direct use, but for a better understanding of the evaluation process and the underlying methodology.

6.2 Virtually Changed Vehicle Geometry

The potential of the methodology introduced here can be illustrated by a virtual experiment: It is well-known from the literature (compare Sect. 5.3.1) that van-like vehicles impose a higher risk of severe injury or fatality to pedestrians than passenger vehicles. Important reasons are a different geometry of the front-end, probable changes in stiffness, and a higher front-end compared to passenger cars. In the virtual experiment, all vehicles in the simulation are given "off-road" capabilities, meaning the vertical dimensions have been virtually enlarged for the whole sample. The lower bumper reference line (see Fig. A.1, p. 179) was used as reference, as increased off-road capabilities or the concept of sport utility vehicles are comparable to a virtual change of the chassis with respect to height. As a consequence, all vertical quantities have been virtually raised by 10 cm, roughly corresponding to one standard deviation of the height of the lower bumper reference line (see Table A.1, p. 182).

Although the whole vehicle fleet was virtually changed, these changes do not influence the processes involved in the situations themselves and thus the accident sample created. The characteristics obtained by the simulation regarding accidents and their genesis are similar to those of the baseline described above.

As explained in Sect. 5.2.6, only a model with geometric quantities among the explanatory factors is able to capture the effects of a changed vehicle geometry.

Table 6.1 gives the results for Option c, i.e., the set of models constructed with the ISS16+ model as starting point. The probability of ISS16+ injuries is about 1.8 % higher in the virtually raised fleet, which is a relative increase of about 28 % compared to the original sample and about 1.3 % higher for ISS9+ injuries (corresponds to a relative increase of 7 %).

This increased risk associated with higher vehicle front-ends is in line with the findings in the literature comparing LTVs and passenger cars (see Sect. 5.3.1) and the direct influence of higher vehicle front-ends as given by the univariate results of Sect. 5.3.1 (independent of the data source used). The injury risk associated with different heights of vehicle front-ends is quantified by odds ratios. Table 6.2 gives the frequency and corresponding odds ratios for the baseline and the baseline with virtual increase in height. The virtual increase in height results in a moderate increase in risk for ISS9+ injuries corresponding to an odds ratio of 1.083, whereas the odds ratio of

Table 6.1 Injury probabilities for the original sample and the virtually raised fleet

	ISS_c					
	0–8	9–15	16–24	25+	9+	16+
Baseline sample (%)	80.5	13.3	3.1	3.1	19.5	6.3
Virtually raised sample (%)	79.2	12.8	4.0	4.0	20.8	8.0

Table 6.2 Injury frequency, odds ratio, and 95 % confidence interval (CI) for the baseline sample and the virtually raised fleet in the simulation

	ISS_c				
	0–8	9+	Odds ratio	95 % CI	
Baseline sample	31,810	7,727	1.083	1.046	1.121
Virtually raised sample	31,536	8,297			
	0–15	16+	Odds ratio	95 % CI	
Baseline sample	37,062	2,475	1.304	1.235	1.377
Virtually raised sample	36,641	3,192			

1.304 for ISS16+ injuries indicates a substantially increased risk for severe injuries. (Due to a large number of accidents available in the simulation, the confidence intervals are rather small.)

In order to compare the simulative results to findings derived from other data sets, comparable odds ratios are constructed using the US PCDS data. The height of the lower end of the front bumper is differently defined in PCDS and GIDAS ($h_{1,PCDS}$ and ($LBRL_{v,GIDAS}$, see Figs. A.1 and A.4, pp. 179 and 181). Thus, the absolute values are not directly comparable. PCDS data include the distinction between LTV and passenger car, GIDAS does not. PCDS data were filtered for frontal impact only, see Sect. 5.2.1, in order to get more comparability to the simulative results. All light truck vehicles, van-like vehicles, and utility vehicles were grouped together as LTVs to increase statistical power (i.e., to obtain as many case numbers as possible). Table 6.3 gives the mean values for cars and LTVs. The difference of the mean bumper bottom height is about 6.4 cm and about 9.8 cm for the bumper top height, which are both highly significant using t-tests ($t = 16.53$, $p < 0.001$ and $t = 9.58$, $p < 0.001$, respectively).

Table 6.3 Bumper bottom and top height for passenger cars and LTVs in PCDS

	Vehicle type	n	Mean	SD
$h_{1,PCDS}$ (cm)	Cars	150	61.29	6.95
	LTVs	300	51.48	3.00
$h_{2,PCDS}$ (cm)	Cars	150	42.98	7.57
	LTVs	300	36.54	4.62

Table 6.4 Injury frequency, odds ratio, and 95 % confidence interval (CI) for PCDS data (Henary et al. [2] and own computations)

	ISS				
Henary et al. [2]	0–16	17+	Odds ratio	95 % CI	
Cars	281	96	1.306	0.879	1.940
LTVs	121	54			
	ISS_c				
PCDS (own computations)	0–8	9+	Odds ratio	95 % CI	
Cars	171	129	1.014	0.682	1.506
LTVs	85	65			
	ISS_c				
PCDS (own computations)	0–15	16+	Odds ratio	95 % CI	
Cars	202	98	1.159	0.768	1.750
LTVs	96	54			

Table 6.4 gives the frequency and odds ratio for comparable injury outcomes. Henary et al. [2] used PCDS data to compute odds ratios regarding injuries and fatalities for LTVs versus passenger cars. Computations based on the data from Henary also indicate an increase in risk for ISS17+ injuries (odds ratio 1.306) for LTVs. A similar trend is indicated by an odds ratio of 1.014 for ISS9+ and 1.159 for ISS16+ injuries when computed directly from PCDS. The PCDS data for own computations were filtered for frontal impact only, which has not been the case in the Henary study.

The confidence intervals for the PCDS data are thus by far larger than the simulative ones, which can be attributed to low case numbers compared to the simulation. The results obtained from simulation are in trend with the results from the literature, such as Henary et al. and with calculations presented here. When more observations become available in accident data in the future, a direct confirmation of the results obtained here is expected.

6.3 Efficacy of Preventive Pedestrian Protection

This section illustrates the results of the process chain for quantitative evaluation of the pre-crash phase as described in Sect. 3.1. A preventive pedestrian protection system is used as example (see Sect. 3.3) together with a simulation as described in Sect. 3.4. In order to explain the methodology behind the process chain, four virtual parameter studies of a preventive pedestrian protection system are performed and analyzed (Table 6.5) using a functional demonstrator:

The system constitutes a system with an optical sensor. The system is assumed to have the same performance independent of any environmental conditions in this

Table 6.5 Definition of four virtual parameter variation studies of a preventive pedestrian protection system

	Study title	System interventions	Parameter variations	Further parameter specifications
1	"Warning"	Driver warning	TTC threshold earliest warning: 1.0–3.8 s	
2	"Warning and brake assist"	Driver warning	TTC threshold earliest warning: 0.8–3.8 s	
		Reconfigured brake assist		Maximum possible automatic deceleration of brake assist: 10.0 m/s^2
3	"Automatic braking"	Automatic braking by the system	TTC threshold earliest automatic braking: 0.4–1.2 s	
			Maximum possible automatic deceleration of automatic braking: 4.0–11.0 m/s^2	
4	"Warning, brake assist, automatic braking"	Driver warning	TTC threshold earliest warning: 0.6–3.8 s	
		Reconfigured brake assist		Maximum possible automatic deceleration of brake assist: 10.0 m/s^2
		Automatic braking by the system		TTC threshold earliest automatic braking: 0.9 s; maximum possible automatic deceleration of automatic braking: 4.5 m/s^2

example (like lighting or weather). The activation parameters (TTC thresholds as well as deceleration capabilities) have been subject of variation whereas vehicle characteristics, system sensor parameters, and, for example, prediction logic of the system have been kept constant (as well as all pedestrian and driver attributes, such as age). The variation is used to show the influence of single system components on the subsequent results of the process chain.

As explained above, the relative difference between variations as well as the absolute efficacy in comparison to the baseline are of interest. Consequently, the virtual experiments include two modes of simulation:

The *first* one creates a baseline, i.e., does not include any preventive system, but creates traffic situations including accidents. Moreover, this simulation not only delivers the baseline, but is also essential for an evaluation of false-positive system actions. To this end, the system is included in an *open-loop* simulation. This means that the system actions are virtual within the simulation, i.e., actually not carried out. For example, if all requirements for a warning are fulfilled, a virtual warning is recorded, but none is actually given to the driver. The number of warnings in the open-loop simulation, prior to non-accident outcomes is a measure of false-positive system actions.

The *second* mode is a *closed-loop* simulation, where all system actions are fully integrated into the driver-vehicle control loop. A simulated warning, for example, is actually given to the driver, or an automatic braking results in a simulated deceleration of the vehicle. This mode allows a quantification of the effects of a particular system configuration on the frequency of accidents and on the resulting injury severities (using the injury probability models of Chap. 5).

The evaluation of false-positive system actions is not possible in a closed-loop simulation, as each system action changes the course of events and influences whether a situation would have resulted in an accident without the system action or not. In the closed-loop simulation, a non-accident situation with system action could thus be either a false positive or an avoided accident.

6.4 Efficacy of System "Warning"

Figures 6.5 and 6.6 give the results for system "Warning" with different thresholds for the earliest possible warning. The baseline consists of the 18 million situations as explained in Sect. 6.1. As the difference in the warning TTC was 0.2 s, each TTC threshold was simulated 100 million times in order to reduce fluctuations in the Monte-Carlo results to a magnitude well below the effect size. All ISS levels in this section have been computed using Option c (starting from ISS16+) for the injury probability models (see Sect. 5.4.1). One would expect as hypothesis that the efficacy for low TTC thresholds converges to zero, as the driver needs a particular time to react, decide on an action, and act in response to a warning.

It can be observed that TTC thresholds between 1.0 and 2.6 s lead to a stronger reduction of accidents and injury levels as larger TTC thresholds. The simulation

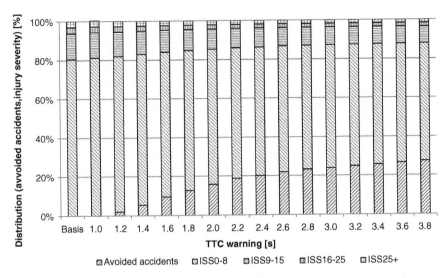

Fig. 6.5 Distribution of pedestrian injury severity and avoided accidents due to system "Warning"

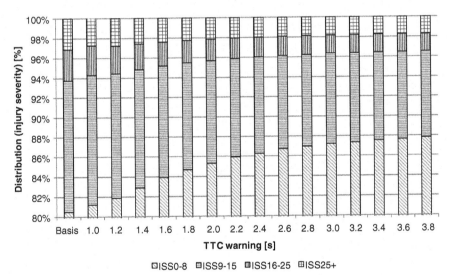

Fig. 6.6 Enlargement of high injury outcome categories of Fig. 6.5

study confirms the hypothesis stated above. Warnings towards a TTC threshold of 1.0 s or smaller evidently have only marginal effects.

For the optimization of a system of active or integral safety, not only the positive effects, as given in Fig. 6.5, are important, but also the overall quality of the system and its components including false positives has to be considered. The number needed to

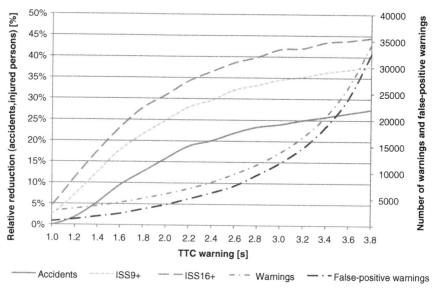

Fig. 6.7 Reduction of accidents and injury severities (*left axis*) as well as warnings (all and false-positive; *right axis*) given to the driver for system "Warning"

treat (NNT) regarding specific outcome metrics is an appropriate metric as discussed in Sect. 2.2 (p. 19).

Figure 6.7 gives the reduction of accidents and injuries relative to the baseline (as ISS25+ is a constant factor relative to ISS16+ in Option c, the relative reduction is identical to ISS16+ and not given in the following graphics). In addition, the number of warnings as well as false-positive warnings issued for each TTC threshold is included. The trends for accidents and injuries have been described above. The number of warnings increases steadily with rising TTC thresholds, with increasing gradient. The earlier a warning is given before a possible accident, the more uncertainty remains in the situation with the pedestrian itself, as more time for avoidance actions by both participants is available. As a result, an increasing number of warnings is given in situations which would not have resulted in accidents and thus are regarded as false-positive warnings.

The number needed to treat (NNT) describes the efficacy of a warning regarding accident avoidance or the avoidance of different injury severities (see Sect. 2.2). The direct relation between warnings and avoided accidents (Fig. 6.8) or injured persons (Fig. 6.9) is quantified by NNT together with the relation of all warnings to true-positive warnings. Regarding accident avoidance, about 17 warnings must be given in the best case to avoid one accident (this is around 2.2 s TTC).

In relation to avoided accidents, the number of warnings shows a stronger increase with increasing TTC thresholds. For decreasing TTC thresholds, the number of avoided accidents decreases more rapidly in relation to the number of warnings.

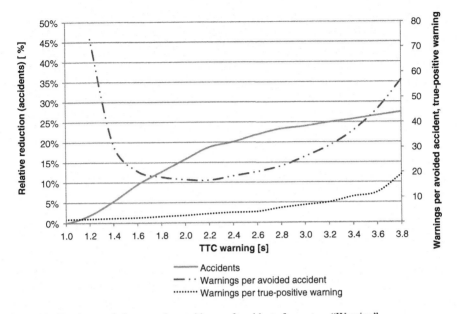

Fig. 6.8 Number needed to treat for avoidance of accidents for system "Warning"

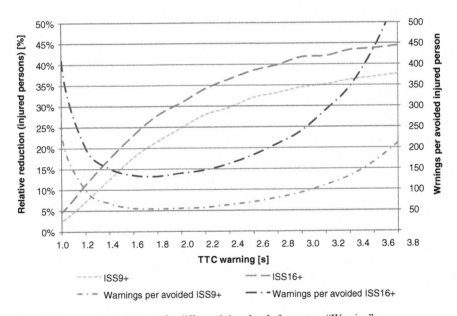

Fig. 6.9 Number needed to treat for different injury levels for system "Warning"

Thus, the NNT shows a U-like shape depending on the TTC of the earliest possible warning. (As there are no avoided accidents for a TTC of 1.0 s, NNT cannot be computed there.) The NNTs for ISS9+ and 16+ injured persons show a similar U-like shape, but the absolute number of NNT is by far greater than that of avoided accidents. If one assumes that avoiding higher levels of injuries justifies higher efforts, also higher absolute values of NNT are acceptable. The fewer warnings required per true-positive warning, the more effective the system is with respect to functional costs. This ratio increases with accelerated pace with increasing TTC thresholds.

Finding an optimal system configuration regarding the TTC threshold for a warning can thus follow several lines. The *first* one sets a goal for accident or injury avoidance and thus uses the reduction given in Figs. 6.5 or 6.7. For example, if the desired objective is an accident avoidance of 20 %, a TTC of about 2.4 s would be appropriate. Consequently, using Fig. 6.8, this would result in an NNT for avoided accidents of 18 and 64 per ISS9+ injury (162 per ISS16+ injury). For every true-positive warning, four false-positive warnings would be given.

Considering these numbers, the developer can decide whether the system quality is sufficient or not. The NNT is especially important if one considers the possible consequences of false-positive system actions. More false activations can lead to lower acceptance or in the worst case to the creation of new critical situations in traffic (see Sect. 2.2). If the consequences of a false-positive warning are assessed using appropriate experiments, a functional "cost function" can be constructed, giving the number of new accidents created by false-positive warnings and inappropriate subsequent reactions of the driver. (An appropriate quantification and the definition of such a function would suggest itself for further research.)

Another approach can directly use NNT in order to find the optimal operating point for the system. In this case, a warning between 1.5 and 2.2 s could be optimal, as all kinds of NNTs for accidents, ISS9+, and ISS16+ have their minimum in that interval. Additionally taking the desired absolute effect of the system into account and considering the consequences of false-positive activations, an operating point can be defined. For example, if the NNTs for accidents and both injury outcomes should be around their minimum and it is desired to avoid about 15 % of accidents, a TTC of 1.9 s could be chosen.

Figure 6.10 gives the *absolute* and *marginal* functional costs depending on the TTC threshold (see Sect. 2.2). The discussion of the U-like shape of overall NNT already showed that an increase at low TTC threshold is beneficial. This is also reflected in the slope of the overall NNT curve as given by the marginal NNT. For example, a change in TTC from 1.0 to 1.2 s for ISS9+ has negative marginal costs (meaning about 150 warning less per avoided ISS9+). A decision to choose a TTC threshold of 1.2 s instead of 1.0 s is thus both beneficial in terms of overall functional costs (about 100 warnings per avoided ISS9+ instead of about 250) and marginal costs. With increasing TTC thresholds, the marginal costs become positive. Each additionally avoided outcome thus is associated with a defined additional effort. If the goals for ISS16+ are maximum overall costs of 400 and maximum marginal costs of 50, the highest TTC acceptable would be 3.2 s (overall costs are about 300 and 3.0 to 3.2 s results in about 50 additional warnings per avoided ISS16+).

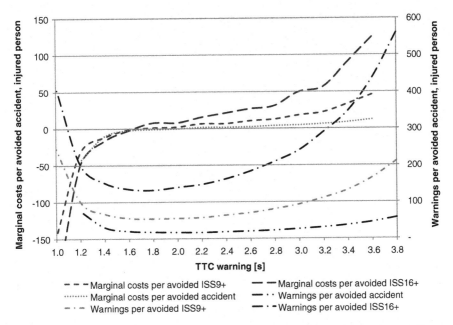

Fig. 6.10 Absolute and marginal number needed to treat for avoidance of accidents and different injury levels for system "Warning". The marginal NNT refers to one incremental increase in TTC

6.5 Efficacy of System "Warning and Brake Assist"

Figures 6.11 and 6.12 give the results for system "Warning and brake assist". The baseline is identical to the system discussed above. Each TTC threshold was simulated in one million crossing situations. As the effect size is greater than for "Warning", one million situations per TTC threshold are sufficient. However, the natural fluctuations of the Monte-Carlo simulation can be observed in the curves for avoided accidents or injuries (see also Fig. 6.13), as those are in theory strictly monotonically increasing.

With regard to Fig. 6.13, the following observations can be made: The higher the TTC threshold, the more accidents and injuries can be avoided. As for "Warning", the decrease is stronger for smaller TTC thresholds than greater ones. The number of warnings and false-positive warnings increases again with an accelerating trend. Reasons and interpretation regarding the warning are similar to system "Warning", as the additional brake assist does not influence activation criteria for the warning itself.

Figures 6.14 and 6.15 give the direct relation between avoided accidents, respectively injured persons, and NNT. The observations are largely comparable to the ones for system "Warning":

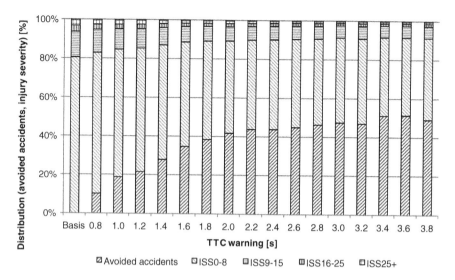

Fig. 6.11 Distribution of pedestrian injury severity and avoided accidents due to system "Warning and brake assist"

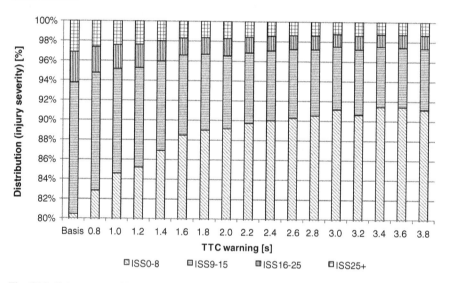

Fig. 6.12 Enlargement of high injury outcome categories of Fig. 6.11

- NNT rises with increasing TTC
- NNT is lowest for avoided accidents, higher for ISS9+, and again higher for ISS16+
- The ratio of warnings per true-positive warnings also shows an increase with increasing TTC

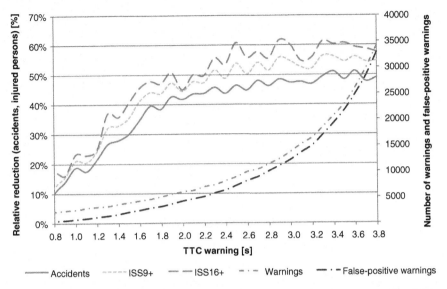

Fig. 6.13 Reduction of accidents and injury severities (*left axis*) as well as warnings (all and false positive; *right axis*) given to the driver for system "Warning and brake assist"

The typical U-shape of NNT can also be observed in Figs. 6.14 and 6.15 (due to an accident avoidance of about 10 % at a TTC threshold of 0.8 s, the left half of the "U" is not that apparent).

At a TTC of 0.8 s, there are already about 10 % avoided accidents, whereas for system "Warning" (see Fig. 6.8), even at a TTC of 1.0 s, there are no avoided accidents. This exemplarily illustrates the additional efficacy created by the reconfigured brake assist. Obviously, the warning has the same effect on the driver regarding catching his attention and triggering subsequent behavioral processes. However, some drivers do see the pedestrian well *before* the warning is issued and are already within their mental process of evaluating, deciding, and initiating an action. In that case, if the warning is issued before the driver hits the braking pedal, the driver does get the 10.0 m/s^2 braking support which is in most cases more deceleration than the "natural" braking without brake assist would have been. As a result, the reason for the effect of the warning at 0.8 s TTC is less the effect of the warning itself, but the brake support by the reconfigured brake assist for already reacting drivers. At even lower thresholds of TTC, the left half of the "U" of the NNT curve would also be visible, as accident and injury avoidance converge to zero. Again, the same consideration regarding functional costs can be made for this example, but are not explicitly discussed here.

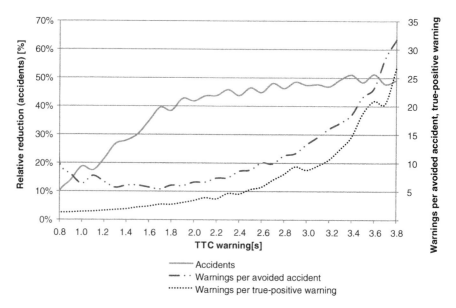

Fig. 6.14 Number needed to treat for avoidance of accidents for system "Warning and brake assist"

6.6 Efficacy of System "Automatic Braking"

Figures 6.16, 6.17, and 6.18 show the efficacy of system "Automatic braking" regarding avoidance of accidents and injuries (the percentages of the baseline are included, e.g., in Fig. 6.11). The efficacy of automatic braking increases with higher TTC thresholds and higher braking decelerations, respectively. (Note, that the surfaces are in theory smooth but show the similar fluctuations, i.e., variance, of the Monte-Carlo simulation with one million situations as described above.)

There are three different approaches to interpreting the surface: *First*, holding deceleration constant, an increase in TTC threshold leads to increasing avoidance. One possible reason is that the earlier the braking TTC, the more time the pedestrian has for escaping the vehicle path and thus resulting in accident avoidance. Another is that the vehicle has more time to brake. However, the accident avoidance is nonlinear (saturation): increases in TTC thresholds do not always imply earlier triggering, since additional conditions such as pedestrian trajectory (with respect to probability of collision) need to be satisfied.

Second, examining lines of constant TTC threshold indicates that maximum braking deceleration has a quite linear effect on accident or injury avoidance.

Third, the contour lines of the surface represent contours of equal efficacy. The gradient to the contours gives indications on the relationship between TTC and deceleration. The gradient at high decelerations is in direction of TTC thresholds; at lower decelerations, the gradient is in direction of both TTC thresholds and decelerations.

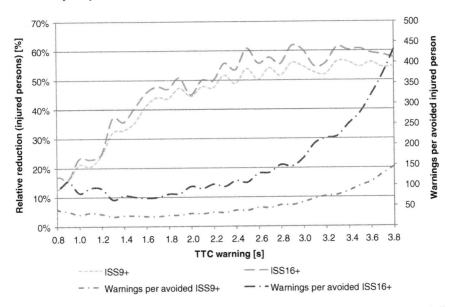

Fig. 6.15 Number needed to treat for different injury levels for system "Warning and brake assist"

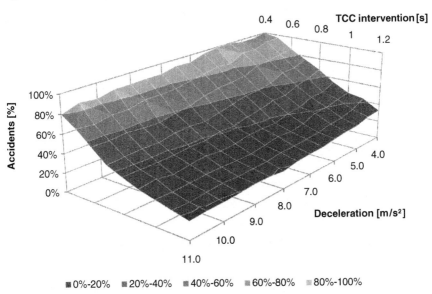

Fig. 6.16 Distribution of accidents for system "Automatic braking"

The relationship is thus not a linear one. At high deceleration, changes in TTC threshold dominate. At lower decelerations, changes in TTC threshold and changes in deceleration are both important. Following a line of equal efficacy, increased TTC thresholds do not lead to linearly decreased levels of deceleration but accelerated

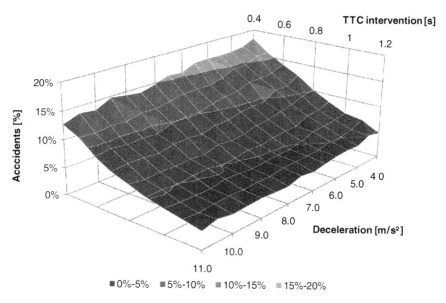

Fig. 6.17 Distribution of accidents with ISS9+ injured pedestrians for system "Automatic braking"

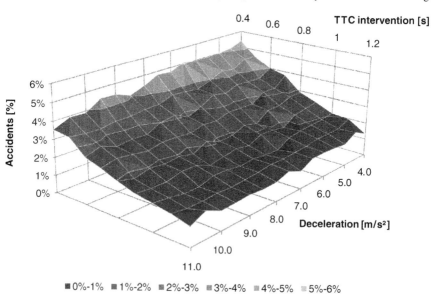

Fig. 6.18 Distribution of accidents with ISS16+ injured pedestrians for system "Automatic braking"

decrease of deceleration thresholds (which can be observed in the change of the direction of the gradient to the contour line).

Figure 6.19 gives the frequency of interventions and Fig. 6.20 the number needed to treat for avoidance of accidents for system "Automatic braking". The number of

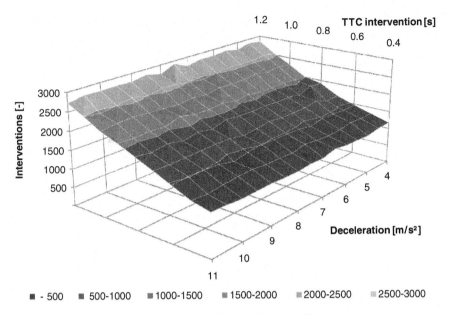

Fig. 6.19 Number of interventions for system "Automatic braking"

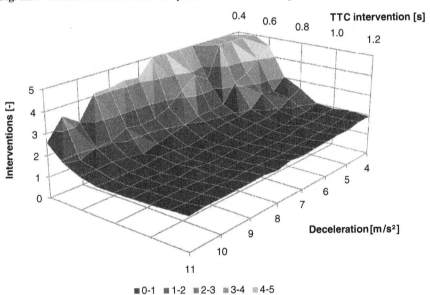

Fig. 6.20 Number needed to treat for avoided accidents for system "Automatic braking"

interventions is obviously independent of the maximum braking deceleration and increases linearly with increasing TTC. The NNT for avoided accidents is basically comparable to the options discussed before. Due to a low number of avoidable

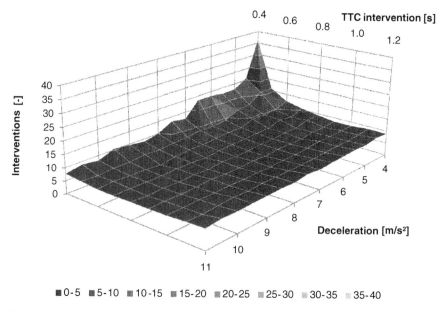

Fig. 6.21 Number needed to treat for avoided ISS9+ injuries for system "Automatic braking"

accidents, the NNT in Fig. 6.20 rises above the maximum of the scale for combinations of low TTC values and low decelerations. The typical U-shape is not prominent, but can still be observed for a variation of TTC values (e.g., at a deceleration of 4.0 m/s^2).

Figures 6.21 and 6.22 give the NNT for avoided injuries for system "Automatic braking". Again, the absolute levels of NNT are higher for ISS9+ than for accidents, and higher for ISS16+ than for 9+. The U-shape is not that clearly visible. An optimization, respectively the definition of an operating point for a system configuration, works as described above (see Sect. 6.4). Especially the optimization using both absolute NNT and the slope of the NNT is discussed on the example in Sect. 6.4.

6.7 Efficacy of System "Warning, Brake Assist, Automatic Braking"

For system "Warning, brake assist, automatic braking" the TTC of the earliest warning was subject to variation, whereas the configuration of the brake assist and the automatic braking was not changed. Again, the fluctuations of the Monte-Carlo simulation with one million situations is visible, as discussed above.

Figures 6.23 and 6.24 give the accident and injury avoidance for system "Warning, brake assist, automatic braking". As all three options discussed above are combined

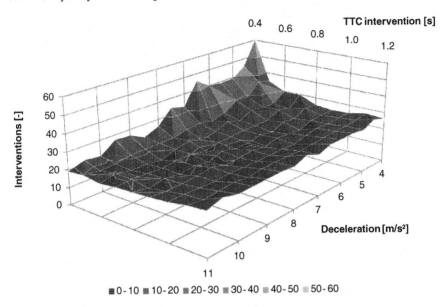

Fig. 6.22 Number needed to treat for avoided ISS16+ injuries for system "Automatic braking"

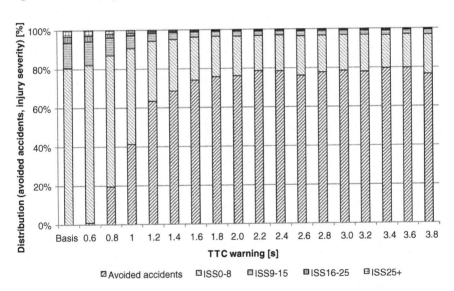

Fig. 6.23 Distribution of pedestrian injury severity and avoided accidents due to system "Warning, brake assist, automatic braking"

here, two main effect are obvious: A very high level of avoidance can be achieved and there is a large increase of avoidance at fairly low thresholds of TTC. The reasons for this are several: The brake support is always effective, once a warning was issued and

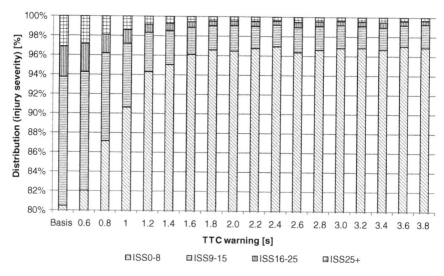

Fig. 6.24 Enlargement of high injury outcome categories of Fig. 6.23

the driver has not already braked. At very low TTC thresholds, where the warning is not very effective, the automatic braking is responsible for the main effect. At higher TTC thresholds, the combined effect of the warning itself (together with the brake assist) and the automatic braking (in case the reaction of the driver was too slow for a significant effect), leads to a high avoidance both in accidents and in injuries. Considering the shape of the curves, the trends are comparable to the system variation as discussed above.

Figure 6.25 gives the relative reduction of accidents and injuries depending on warning TTC together with the absolute number of warnings, false-positive warnings, and intervention. The reduction of accidents and injuries follow a sharp increase at low warning TTCs (between 0.6 and 1.2 s warning TTC). Above 1.5 s warning TTC the reduction becomes steady and then marginal. The warnings show a similar increasing trend with increasing warning TTC.

The number of interventions is quite steady around 1,623 interventions with a standard deviation of 233, showing an increase at low warning TTCs and a maximum of 2,015 interventions at a warning TTC threshold of 1.5 s. With increasing warning TTC, the rate of decrease of the number of interventions decreases. At high warning TTCs, the number of situations which have become critical enough to fulfill the activation criteria for an intervention is quite constant. It resembles a rather constant percentage of situations, which cannot be handled appropriately by the driver (with or without a warning). At low warning TTCs, driver reactions may be too late or insufficient in more cases than at high TTC thresholds, so the criteria for activation for automatic braking are fulfilled more frequently.

The number needed to treat for accident and injury avoidance are given in Figs. 6.26 and 6.27. As system "Warning, brake assist, automatic braking" includes

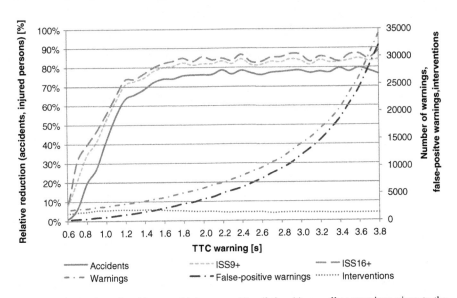

Fig. 6.25 Reduction of accidents and injury severities (*left axis*) as well as warnings given to the driver and interventions (*right axis*) for system "Warning, brake assist, automatic braking"

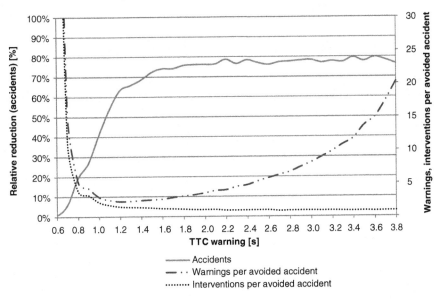

Fig. 6.26 Number needed to treat for avoidance of accidents for system "Warning, brake assist, automatic braking"

both warning and intervention, both NNTs have to be considered together. It can be observed that the NNT for warnings shows the typical U-like shape for all outcome categories. The NNT for interventions on the contrary has an L-like shape. Towards

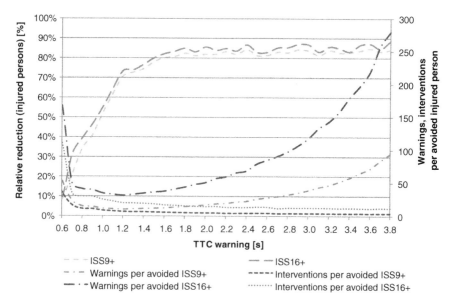

Fig. 6.27 Number needed to treat for different injury levels for system "Warning, brake assist, automatic braking"

low warning TTCs, the system is not able to avoid many accidents or injuries. Thus, the NNT becomes very large. On the other end (at high TTCs), both the number of interventions as well as the number of accidents or injuries stay rather constant.

Although the U-like shape of the warning NNTs allow for an easy determination of the optimum range with respect to the parameter in variation (i.e., warning TTC in this example), the L-like shape of the intervention NNT requires a more subtle interpretation in order to find an optimum. The question is how both NNTs have to be interpreted in combination to find the optimal strategy and the operating point for the whole system. It is evident that a quantity including both of these characteristics is needed.

6.8 Comparison of Warning and Intervention

The different shapes of the NNT curves for warnings and interventions complicate the search for an operating point. One possible solution is the definition of a common scale for both warnings and interventions. Warnings and interventions produce similar effects in the end (i.e., accident or injury avoidance), but induce totally different processes in traffic. A warning in this case is directed towards the driver and triggers the following process: The driver searches for the reason of the warning. Once he finds out (i.e., the critical pedestrian situation), he evaluates the situation, decides, and eventually initiates an action (in this case braking). In the case of automatic braking, the vehicle acts without the driver and initiates a predefined braking maneuver.

For a true-positive system action, both variants are different only in their efficacy. However, the consequences of a false-positive action are considered different regarding their severity: A warning may eventually lead to a braking by the driver, but the intensity and duration of the braking is up to the driver. The driver's evaluation and decision loop additionally helps to avoid braking maneuvers as consequence of false-positive warnings. A false-positive automatic braking lacks this second evaluation of the situation.

Depending on the deceleration and the duration of the braking (by both the driver and the vehicle itself), undesired side effects in the upstream traffic stream could be the consequence. If a braking maneuver includes high decelerations and cannot really be anticipated by the following traffic, rear-end collisions could be the result.

It is well known that false-positive automatic braking maneuvers become more critical for upstream traffic with increasing deceleration and duration (i.e., absolute velocity decrease and time). For example, whereas a braking with 4.0 m/s^2 is commonly regarded as uncritical, braking with very high deceleration (e.g., 10.0 m/s^2) is considered far more critical by many experts regarding controllability. The driver could be shocked by a false-positive intervention involving a high deceleration and may not react properly. Warnings instead are believed to induce fewer unnecessary braking maneuvers, as the driver is still in control and has the chance to evaluate the situation before initiating a maneuver.

Using system "Warning, brake assist, automatic braking" as example, the NNT using effective interventions (see Sect. 2.2, p. 19) is given in Fig. 6.28. The factor comparing the consequences of warnings and interventions is arbitrarily set to 10 for this example, meaning an intervention with 4.5 m/s^2 has the same effect in traffic as 10 warnings given to the driver. The NNT with effective interventions is thus a weighted combination of the NNTs for warnings and interventions. It provides a single criterion and thus allows a systematic optimization procedure even in the presence of multiple control parameters (e.g., TTC thresholds, braking deceleration). For avoided accidents, an optimum for NNT can be found in this case around a warning TTC of 2.0 s. The optima for ISS9+ and ISS16+ are not that easily visible, but can be expected in the same region, as all three kinds of NNT for warnings show similar trends in Figs. 6.26 and 6.27 (the corresponding trends for interventions are also similar). Of course, the concept of marginal NNT can be applied as well.

Again, the actual factor between warnings and interventions depends on the specific design of the warning as well as the parameters of the automatic braking. Both have to be evaluated in dedicated studies in order to determine the factor necessary for a meaningful combination into effective interventions. This is one objective of ongoing research in the field of controllability.

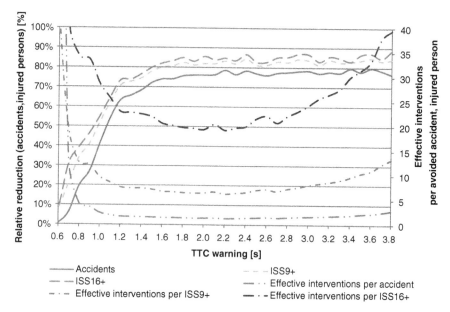

Fig. 6.28 Reduction of accidents and injury severities (*left axis*) as well as effective interventions (*right axis*) given to the driver for system "Warning, brake assist, automatic braking"

6.9 Conclusion

The practical use of the processes and methods for evaluating active and integral safety described in the chapters above have been illustrated and discussed in this chapter. Preventive pedestrian protection was prospectively evaluated using a stochastic simulation of potentially critical traffic situations. The efficacy was compared for different system variations including variation of key system parameters. The application of the injury probability models developed in Chap. 5 as well as the interpretation of the results were explained regarding the definition of an operating point.

The reference or baseline for virtual experiments evaluating different configurations of a preventive pedestrian protection system is a simulation sequence without any system. The simulation generates a high number of traffic situations in which a pedestrian is going to cross a street. The scenario considered here results in a low percentage (about 0.2 %) of accidents compared to all situations simulated. As the accident fraction is a result of many subprocesses modeled (which themselves have their particular means of validation) the validation of this fraction of situations gives an indication on the plausibility of the whole modeling. The characteristics of the accidents can be made plausible or can be validated using existing accident data bases. A corresponding comparison of the accident fraction using GIDAS data was given in the first section of this chapter.

Since one strength of the injury probability models developed in Chap. 5 are the explanatory factors included (e.g., pedestrian and vehicle attributes besides crash characteristics), a virtual experiment was conducted to show the plausibility of the simulated results and investigate the effects of hypothetical geometric changes in the vehicle fleet. The vehicle fleet was virtually raised by 10.0 cm and thus given the geometric characteristic of light truck vehicles. This leads to an increased risk for different injury levels for the pedestrian, which is in line with previous findings from literature.

The next section evaluates a variety of different preventive pedestrian protection systems using a functional demonstrator. Three possible system components have been investigated: warning, warning in combination with a reconfigured brake assist, and automatic braking. Variations of different key system parameters, e.g., earliest TTC of activation or desired braking deceleration, have been evaluated. The efficacy is illustrated by the change in avoided accidents as well as changes in the injury distribution as given by ISS. The functional "costs" of each system variant are quantified by the absolute number of system actions as well as the number needed to treat, i.e., the ratio of system actions to avoided accidents or injuries. Different ways for system development and optimization have been introduced: Starting point can be either a desired efficacy (both in avoided accidents or injuries of a specific level) or the absolute and/or marginal NNT acceptable. Both approaches will allow the determination of an optimized system strategy.

Finally, a system option combining all three components was evaluated. The overall efficacy is achieved by warnings, the reconfigured brake assist, as well as the automatic braking. The different nature of warnings and automatic braking with respect to the upcoming traffic is described and a solution for system optimization is introduced. As the NNT for warnings follows a characteristic U-like shape (which makes the determination of a minimum possible), the NNT for interventions produces a L-like shape. With the introduction of the theoretical quantity of effective interventions, a factor for the direct comparison of warnings and interventions is created. Warnings and interventions are combined in a weighted sum. The weight represents the risk in traffic associated with each class of system action. The actual quantification for a specific warning and intervention concept has to be determined by targeted research (emphasizing controllability). The effective interventions can again be computed into an NNT, and the optimal operating point with respect to the system parameters in variation can be found.

The concept of marginal functional costs creates the basis for an incremental search for the optimal system configuration. Once the range of acceptable overall functional costs (given by NNT and, if necessary, by effective interventions) is found, the additional functional costs per increment of the optimization parameter is quantified by the marginals. This allows for a more targeted and specific development of measures with respect to the effort accepted by the stakeholders in charge and the desired outcome.

A process for the evaluation of active and integral safety has been explained from the concept of a process chain, the description of a simulative method, the development of traffic scenarios, the construction of injury probability models, and the explanation of the whole methodology using the example of preventive pedestrian protection.

References

1. GIDAS German In-Depth Accident Study. Retrieved December 22, 2010, from http://www.gidas.org/files/GIDAS_eng.pdf.
2. Henary, B. Y., Crandall, J., Bhalla, K., Mock, C. N., & Roudsari, B. S. (2003). Child and adult pedestrian impact: The influence of vehicle type on injury severity. *47th Annual Conference of the Association for the Advancement of Automotive Medicine.*

Chapter 7
Conclusion and Outlook

The history of vehicle safety is to a large extent the history of passive vehicle safety. Accident research, the laboratory testing of passive safety by means of crash tests, and subsequent development and improvement of technical measures for mitigation of injuries once an accident has occurred, have dominated vehicle safety for decades. Advances in technology, especially in electronics and computing, have led to the genesis and extensive implementation of driver assistance functions in vehicles, both for safety and comfort. For safety assessment, this means a change of paradigm. The aim of mitigating the consequences of an accident (i.e., passive safety) is increasingly being combined with the aim of avoiding the accident entirely (i.e., active safety).

Trends in accident statistics over the last decades reveal (Chap. 1) impressive improvements in vehicle safety due to a combination of the three E's of traffic safety: engineering (both vehicle- and infrastructure-based), education, and enforcement. Overall changes due to active or passive safety are easily assessed using accident statistics: Fatalities and injury frequencies per accident in vehicles give an indication of improvements in passive safety; effects due to active safety are evident using the ratio of accidents to exposure, e.g., distance driven. Improvements within active safety because of a specific system are more difficult to evaluate using accident data: avoided accidents do not directly enter the statistics any more and mitigating effects are sometimes hard to distinguish, i.e., masked within the data, e.g., due to simultaneous improvements in passive safety or other relevant aspects in the traffic system.

The testing and evaluation of vehicle-based safety has been a standardized process for passive safety during the last decades. For active safety, testing procedures and evaluation methods have become standard only in the field of autonomous stabilization of the vehicle. More recent functions, such as rear-end collision warnings, are still subject to a variety of evaluation schemes. Standardization in terms of methods, tools, and procedures has started only recently and will take probably years for final harmonization.

This thesis has focused on the development of a methodology for representative and reliable evaluation of active safety. The practical example studied was preventive pedestrian protection. Active safety systems act within a complex, dynamic

© Springer International Publishing Switzerland 2015   171
T. Helmer, *Development of a Methodology for the Evaluation of Active Safety using the Example of Preventive Pedestrian Protection*, Springer Theses,
DOI 10.1007/978-3-319-12889-4_7

traffic environment; thus, a feasible and reliable process for evaluation including a stochastic simulation of traffic was defined. The aim was to predict the contribution of an active safety system to reduction of mortality and injuries as well as possible negative consequences induced by unintended system actions, such as false-positive activations.

The introduction included basic models essential for an understanding of traffic and accident genesis. A summary of accident statistics of Germany and the US with respect to pedestrians and the overall situation and recent trends were given; accidents have to be regarded as statistically rare events and a kind of "anomaly" in traffic. The insights on pedestrian accidents, derived from the accident statistics, were then entered into a top-down model for deriving functions and systems capable of addressing the problem, i.e., fulfilling the vehicle characteristic "vehicle safety" with respect to pedestrians. A short summary of recent regulations and technical approaches (all passive safety) defined the state of the art in vehicle-based pedestrian protection.

A review of the current state of scientific and technical knowledge on evaluation of the pre-crash phase set the starting point for this thesis (Chap. 2). Safety evaluation can be conducted at different levels (e.g., component-, system-, vehicle-based or with focus on the overall benefits in traffic). The method of choice depends on the level of evaluation and the underlying research question. Functions of active safety rely on sensors which perceive information from their environment and are thus subject to uncertainty. Besides possible technical limitations, the prediction of future movements of all involved participants contributes to this inherent uncertainty. As a consequence, systems subject to uncertainties will not work perfectly in the sense of reliability. False-positive activations, e.g., due to misinterpretation of information or technical limitations, will occur; with consequences on acceptance and controllability of the system. With an increasing number of false-positive activations, acceptance by the driver will decrease. In case of severe interventions in traffic, such as high velocity reductions and sharp decelerations, false-positive activations become a matter of controllability for the driver and the surrounding traffic and can ultimately have a negative impact on safety.

Not only the specific function or system of active safety, but also vehicle characteristics (such as driving dynamics), traffic itself, and especially human behavior, are important elements in both accident genesis as well as avoidance. As a consequence, an evaluation method must include all relevant elements with their specific distributions. A particular challenge to assessment is the huge combination of possible situations and the ability to produce meaningful and representative results with respect to the traffic situation in question. Furthermore, the evaluation method must be capable of predicting future effects (prospective approach) and thus not only assess developments of the past (retrospective approach). Since detailed knowledge on all these elements is necessary during evaluation, the spectrum of common methods, procedures, and tools was introduced and discussed. Some of them, e.g., FOTs, do not only provide valuable input for modeling but also enable validation of various aspects, especially regarding critical traffic situations.

The ideal quantification of safety changes due to active safety would provide direct estimation of mortality and injury reduction from accident statistics and direct measurement of false positive counts in the field. However, estimation of ADAS safety benefits from accident statistics requires long observation periods and is confounded by multiple parallel influences on these statistics; false-positive rates need to be measured not just once, but for each algorithmic threshold setting. Hence, a methodology is required that can predict mortality and injury reduction as well as false positives.

An evaluation that addresses the overall safety benefit of a measure in a given traffic system (e.g., a country) must thus consider both positive and negative effects. Existing schemes and methods for evaluating safety functions were reviewed regarding their ability to assess overall safety benefits. Nearly all methods available (many of them involving simulation techniques) focus on the safety effects in a sample of existing accidents. There, essentially the positive effects can be assessed, as the majority of negative consequences mainly take place in traffic situations that would not have led to an accident.

As a possible positive safety effect is evident mainly within accidents (instead of non-accident situations), common approaches rely on reconstructed accident data and simulate the effect of an active safety system. However, there are several well-known limitations: False-positive system actions (and consequently an important component of overall functional "costs") cannot be adequately assessed, as no representative sample of situations in which the system would be triggered (including non-accident situations) can be generated. Also assessment based on accidents can be sensitive to details of the accident reconstruction, which are indeed subject to uncertainties. However, a particular instance of a reconstructed accident may not be entirely representative, particularly regarding the effectiveness of a proposed assistance system.

The new approach and statistical analyses presented in this thesis (Chap. 3) provide a path for the evaluation of active safety with respect to its safety impact on a traffic system. The rising need to answer the question of overall safety effect in traffic could thus be met in a representative and statistically stable way. The many ways in which uncertainty enters an approach and leads to variability in the results, can be addressed by the process presented and quantified by confidence intervals. Requirements for the method and an assessment process including data sources, modeling, simulation, and evaluation were defined. The starting point for the development and testing of a function was an understanding of the safety problem. To this end, reference scenarios for pedestrians (derived from accident statistics) were used. The most important pedestrian scenario for Germany and the US is the "mid-block dash" (i.e., vehicle going straight, pedestrian crossing). A functional demonstrator of a system of active safety was defined to address this situation. The system strategy can involve warnings to the driver, enhanced brake assist, and an automatic braking maneuver.

The traffic situation leading to the described pedestrian accident scenario (i.e., a pedestrian trying to cross a straight road) was modeled in a traffic simulation including all relevant parameters with their realistic distribution in terms of exposure.

The vast majority of situations in the simulation did not result in an accident. The characteristics of simulated accidents as a "random" result were tested using knowledge from in-depth accident studies such as GIDAS. The simulation used showed adequate validity.

The impact of avoidance and mitigation of pedestrian collisions can be evaluated from individual (i.e., physical or physiological), societal, or economic points of view. Different effectiveness measures with their advantages and challenges were discussed. The key aspect for assessment for all of these points of view is a metric that quantifies reduction of injuries and their severity. In principal, crash simulations could be used to predict injury distributions. However, detailed crash simulations (as common in passive safety) require high computational resources and are currently only an option for a small selection of cases. Hence, due to the high number of simulation runs necessary for statistical significance of the results, a less computationally intensive method was required for estimating the probability distributions of physiological outcomes from physical quantities. To this end, this problem was solved using detailed probabilistic models as presented above.

As a simulation of the effects of preventive pedestrian protection in a given traffic situation required considerable model input, a driving simulator study was conducted (Chap. 4). Driver behavior with respect to acceptance of such a system, especially during false-positive actions, was assessed. False-positive system actions were less acceptable for the driver if the pedestrian was not perceived as endangered. If a false-positive system action was unpredictable for the driver (e.g., no pedestrian could be seen), the vehicle had relatively higher speed, or the driver was carrying out a complex maneuver, the system action was rated potentially hazardous for surrounding traffic. High attention of the driver thus decreased the perceived level of hazard in such a situation.

Another aspect of the driving simulator study was the investigation of uncritical interactions with pedestrians. Normally, pedestrians were passed by at an average lateral distance of 1.5 m, and the subjects started braking at an average TTC of 4 s. These findings give valuable input for the design of a system. If the system acts well within these limits, acceptance can be assumed to be high if the driver has not reacted himself in advance. The crucial point with system acceptance and safety benefit is that reduced acceptance will trigger deactivation of the system (if possible) and thus reduce the safety benefit to zero. If the system is optional equipment, a driver with very low acceptance is likely not to include it in his next vehicle, resulting also in zero safety benefit.

An additional finding of the study was that a realistic investigation of highly critical situations proves to be challenging in a driving simulator, since most drivers do not experience an accident, even without the system (baseline). Despite optimized experimental design and additional distraction by a tested secondary task, the drivers were able to perceive the hazards quite early. Possible reasons as well as solutions were discussed together with the advantages and limitations of driving simulator studies in this context.

A main focus of the thesis was the construction of injury probability models for the pedestrian in frontal vehicle crashes (Chap. 5). The literature review conducted resulted only in few models and revealed many open research questions. To address these, the aim of this part was the construction of probability models with respect to the outcome category (ISS versus MAIS), the number of explanatory factors to be included in the models (multivariate vs. univariate), and the modeling of specific age groups (one model for all ages versus different models for different age groups). A new approach of constructing probability models for several cumulative outcome categories by means of conditional probabilities was developed.

Probability models for pedestrians regarding different injury levels as well as fatalities in frontal vehicle crashes were estimated using both German (GIDAS) and US (PCDS) in-depth accident data. Data preparation steps including consistency checks, data scaling, and especially detailed procedures for imputation of missing data were key requirements for utilizing these data sets. A procedure for quantifying the variance associated with imputation was developed and implemented. Recoding and transformation were introduced in order to support comparability of odds ratios associated with different distinct factors, as obtained by logistic regression. Consequently, the effect sizes were comparable between different quantities, as all continuous variables had been standardized.

The resulting models and their validity were analyzed: In-sample predictive accuracy was assessed via the area under the curve (AUC) of the receiver operator characteristic (ROC). The expected out-of-sample predictive accuracy was quantified by 10-fold cross-validation, with the aim of ensuring high validity and at the same time avoiding over-fitting. Challenges when using observational data, such as multicollinearity, confounding factors in analysis, and selection effects were addressed and accounted for.

With regard to the research hypotheses, the following results were observed. There was a clear trend that the Injury Severity Score (ISS) has advantages over the Maximum of the Abbreviated Injury Scale (MAIS) as target variable. Multivariate models seemed to be more accurate than univariate ones, although the differences were not significant for every model regarding ROC AUC, which is presumably due to low case numbers. The statistical power of the sample available here for investigating the use of age specific injury models was analyzed and was found to be too low. A possible future approach toward obtaining detailed insights regarding injury severity and distribution with relation to particular age groups could be detailed crash experiments, either virtual or real. These kind of detailed investigations could be very important for future system designs.

The general findings of the models, the contained factors, and their effect size were in line with previous results in the literature: for example, impact speed of the vehicle was by far the most important predictor for both injury severity and mortality, in both data sets and all models. Pedestrian age was also a key predictor. Confirming a longstanding hypothesis in the literature, the different models obtained here quantified the effects of vehicle profile characteristics and pedestrian attributes (such as BMI).

Two constraints were incorporated into the models: First, zero vehicle speed results in zero injury probability. Secondly, the probability for a more severe (i.e., higher) cumulative target category must not be greater than for a less severe cumulative category. While the first constraint is a definition, the second can be violated if not taken into account while modeling different injury levels. Using a conditional probability simulation generating a very high number of different combinations of all explanatory factors contained in the models, the second constraint was tested; some models, especially with "close" outcome categories, violated the constraints.

This second constraint was inherent within a new approach for constructing probability models for several cumulative outcome categories, e.g., ISS0-8, ISS9-15, and ISS16+, by means of conditional probabilities. These new models generically fulfill the second constraint. This new method also allows for different ways of constructing the models. The level of modeling detail is of course limited by the overall power of the sample. However, it allows the assessment of multiple cumulative outcome levels at once regardless of the number of levels or explanatory factors included in the models.

The accuracy as well as the power of the models depend on the number of cases available; resulting practical limitations to research were highlighted using the GIDAS data base as an example. Especially for the construction of injury probability models, data sources should be up to date and should include far more cases. In addition, the accuracy of the models depend on the quality of the data used; continuous improvements in coding and reconstruction are thus strongly encouraged. Imputation procedures, as included, for example, in US accident data bases, could minimize loss of data due to list-wise deletion in a standardized way.

Stochastic simulation was used together with the probabilistic models to give an example of the application of the assessment process as a whole (Chap. 6). The virtual sample of accidents in the baseline was validated with GIDAS data of corresponding accidents. The multivariate logistic regression model used also contained geometric vehicle characteristics. The baseline simulation with a virtually raised vehicle fleet resulted in changes in the injury severity distribution comparable to well-known findings in the literature, especially regarding light truck vehicles: Higher vehicle front-ends increase the risk of severe injury.

Four different variants of a preventive pedestrian protection system were evaluated regarding their efficacy compared to the baseline using a functional demonstrator. The system used is regarded as a virtual prototype, but nevertheless resembles a realistic system in every aspect and element, e.g., sensor, algorithm, actuator. Each of the basic system settings (i.e., "Warning", "Warning and brake assist", "Automatic braking", and "Warning, brake assist, automatic braking") was subject to variation of at least one optimization parameter. The findings were interpreted with respect to the following metrics: avoided accidents, mitigated accidents (by several injury severity levels) and number needed to treat (resembling the functional "costs" in system actions per desired outcome category). The new concept of effective interventions, as introduced here, combines the functional "costs" of both warnings and automatic system interventions using a hypothetical weighting factor into one key parameter. This "cost function" allowed the direct comparison of systems including only warning or automatic interventions or both. The optimization process regarding an ideal operating point was illustrated. In addition to the absolute priority of relative reduction in avoided accidents or avoided injuries, the objective function for optimization can include absolute functional system "costs" (including total number needed to treat) as well as a target for marginal cost/benefit. However, optimization of a system using the methodology presented is not limited to variations of algorithmic thresholds, but can also include key characteristics of sensors, algorithm, or other relevant vehicle functions.

The most striking improvement provided by the developed methodology is the inclusion of the traffic system as a whole (including accidents) into the evaluation. This approach has incorporated identification of target scenarios; calibration and validation of stochastic behavior (both of technical and human aspects) and injury probability models; stochastic (Monte-Carlo) simulation of target scenarios in varied traffic contexts with/without active safety; quantification of simulative results by appropriate metrics; and integration of supporting and corroborating field and laboratory analyses. This generic assessment process has been demonstrated using the example of preventive pedestrian protection but can be applied to various problems of active safety. For example, the conditional probability approach for ISS level classification is also applicable to other accident victims, such as cyclists or vehicle occupants. Analogous issues of uncertainty regarding sensors and human action also arise in the context of cyclist protection or other constellations, such as vehicle–vehicle interactions, and could also utilize stochastic simulation.

The modeling of technical components (e.g., sensors) with their stochastic elements and intrinsic physical uncertainties, as well as whole systems within a vehicle will continue to be a challenge for the future. One important aspect will be the construction of test benches, where actual hardware can be assessed in a representative way and the results can be used to parameterize the corresponding models. Further research is thus encouraged in order to assess driver behavior under various circumstances, including population-specific behavior characteristics representative for the population of a country. In this context, combined further research on mistakes, conflicts, and accidents in traffic is also strongly encouraged. A special emphasis on the controllability of specific situations by the driver and surrounding traffic regarding false-positive system actions and their consequences will be focus of further research. As the method developed here is applied to future systems, there will be a growing need for targeted investigations of all aspects of the driver-vehicle-environment control loop, including the behavior of drivers and other road users under specific conditions.

Stakeholders in traffic safety require a balanced and comprehensive assessment including positive as well as possible negative effects. The methodology used for optimization is applicable by all stakeholders in development, deployment, and usage of assistance systems, e.g., society, legislators, consumer protection advocates, manufacturers, and customers. Depending on the stakeholder's point of view, utility and generalized costs (negative utility) associated with advanced driver assistance systems can arise from several sources: Decreased acceptance by either driver or society is an example of a generalized cost (negative utility). System development costs include, e.g., definition of operating parameters, design of an optimized combination of active and passive safety measures, etc. Utility arises from reduction of direct monetary costs from, e.g., insurance and health-care, or from reduced indirect monetary costs, e.g., losses of productivity and quality of life. The optimal design of an advanced driver assistance system in principal requires consideration of all these contributions to utility and costs. The concept of assessment introduced here supports objective economic decisions including costs and benefits and thus provides a basis for solving the complex problem of determining the operating point for an advanced driver assistance system.

Appendix A

See Figs. A.1, A.2, A.3, A.4, A.5 and A.6 and Tables A.1, A.2, A.3 and A.4.

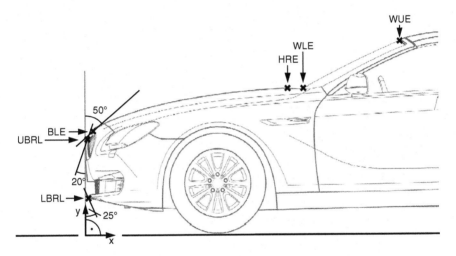

Fig. A.1 Definition of characteristic points at the vehicle front end used in connection to the GIDAS data set

© Springer International Publishing Switzerland 2015
T. Helmer, *Development of a Methodology for the Evaluation of Active Safety using the Example of Preventive Pedestrian Protection*, Springer Theses,
DOI 10.1007/978-3-319-12889-4

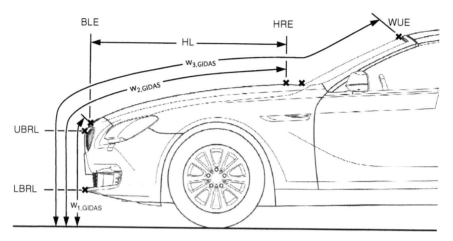

Fig. A.2 Vehicle profile characteristics used for analysis (GIDAS): wrapping distances

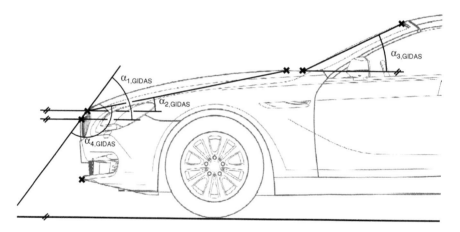

Fig. A.3 Vehicle profile characteristics used for analysis (GIDAS): definition of characteristic angles

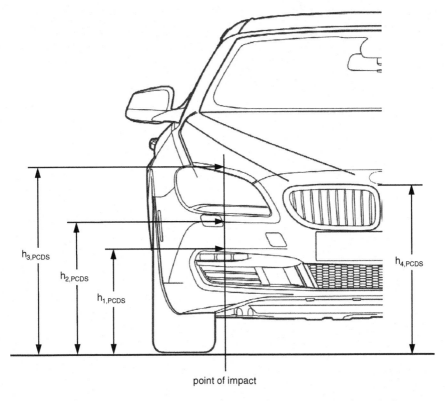

point of impact

Fig. A.4 Vehicle profile characteristics used for analysis (PCDS): characteristic vertical measurements

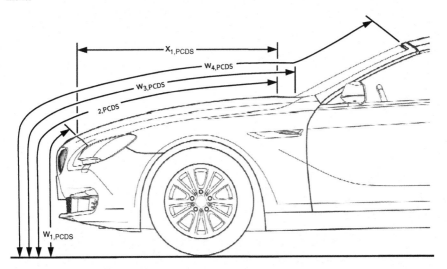

Fig. A.5 Vehicle profile characteristics used for analysis (PCDS): characteristic angles

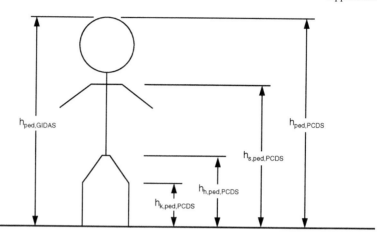

Fig. A.6 Pedestrian vertical measurements in GIDAS and PCDS

Table A.1 Continuous variables used (GIDAS)

Variable name	Symbol	Unit	N	Mean	SD
Vehicle kinematics					
Impact speed	$v_{c,GIDAS}$	kph	930	29.35	17.04
Impact speed (squared)	$v_{c,GIDAS}^2$	(kph)^2	930	1151.60	1473.47
Kinetic energy	E_{kin}	kJ	877	738.45	1008.52
Mean braking deceleration before collision	a_{veh}	m/s^2	822	38.47	33.31
Vehicle characteristics					
Crash weight	$m_{veh,GIDAS}$	kg	933	1264.15	273.18
Year of first registration	y_{veh}	a	937	1995.69	5.10
Lower bumper reference line—longitudinal	$LBRL_{l,GIDAS}$	cm	906	3.84	2.85
Lower bumper reference line—vertical	$LBRL_{v,GIDAS}$	cm	896	29.99	9.16
Upper bumper reference line—longitudinal	$UBRL_{l,GIDAS}$	cm	906	0.41	0.61
Upper bumper reference line—vertical	$UBRL_{v,GIDAS}$	cm	896	51.93	4.05
Height of front bumper	$HFB_{v,GIDAS}$	cm	906	51.44	5.33
Bonnet leading edge—longitudinal	$BLE_{l,GIDAS}$	cm	906	12.40	3.02
Bonnet leading edge—vertical	$BLE_{v,GIDAS}$	cm	896	73.95	6.53
Bonnet leading edge (wrapping distance)	$W_{1,GIDAS}$	cm	896	77.21	6.62
Hood rear end—longitudinal	$HRE_{l,GIDAS}$	cm	906	101.54	20.81
Hood rear end—vertical	$HRE_{v,GIDAS}$	cm	896	95.46	6.88

(continued)

Table A.1 (continued)

Variable name	Symbol	Unit	N	Mean	SD
Hood rear end (wrapping distance)	$W_{2,GIDAS}$	cm	896	170.15	16.09
Hood length	HL_{GIDAS}	cm	896	92.94	16.53
Windshield upper edge—longitudinal	$WUE_{l,GIDAS}$	cm	906	173.98	24.99
Windshield upper edge—vertical	$WUE_{v,GIDAS}$	cm	896	134.94	9.65
Windshield upper edge (wrapping distance)	$W_{3,GIDAS}$	cm	896	253.54	13.53
Angle of upper bumper ref. line to bonnet leading edge	$\alpha_{1,GIDAS}$	°	906	61.25	7.04
Angle of hood	$\alpha_{2,GIDAS}$	°	896	14.35	6.24
Angle of windshield	$\alpha_{3,GIDAS}$	°	896	31.15	3.45
Angle of bonnet leading edge	$\alpha_{4,GIDAS}$	°	896	137.35	8.33
Pedestrian physiology					
Age	y_{ped}	a	1011	35.91	25.83
Body weight	$m_{ped,GIDAS}$	kg	1004	60.99	21.35
Body height	$h_{ped,GIDAS}$	cm	1006	160.27	20.26
Body mass index	BMI	kg/m^2	1004	22.87	5.21
Height to upper bumper reference line—vertical	$r_{1,GIDAS}$	–	889	3.10	0.44
Height to bonnet leading edge—vertical	$r_{2,GIDAS}$	–	889	2.18	0.32
Height to hood rear end—vertical	$r_{4,GIDAS}$	–	889	1.69	0.24
Height to windshield upper egde—vertical	$r_{6,GIDAS}$	–	889	1.19	0.17
Height to bonnet leading edge (wrap)	$r_{3,GIDAS}$	–	889	2.09	0.30
Height to hood rear end (wrap)	$r_{5,GIDAS}$	–	889	0.95	00.16
Height to windshield upper edge (wrap)	$r_{7,GIDAS}$	–	889	0.63	0.09

Case numbers (N), mean, and standard deviation (SD) are given for the full data set

Table A.2 Continuous variables used (PCDS)

Variable name	Symbol	Unit	N	Mean	SD
Vehicle kinematics					
Impact speed	$v_{c,PCDS}$	kph	376	28.95	20.80
Impact speed (squared)	$v_{c,PCDS}^2$	$(kph)^2$	376	–	–
Kinetic energy	E_{kin}	kJ	376	69.99	113.14
Vehicle characteristics					
Vehicle curb weight	$m_{veh,PCDS}$	kg	450	1415.10	340.59
Front bumper bottom height	$h_{1,PCDS}$	cm	450	38.68	6.52
Front bumper top height	$h_{2,PCDS}$	cm	450	54.75	6.59
Front bumper lead	$x_{1,PCDS}$	cm	449	9.31	2.96
Angle of front bumper	$\alpha_{1,PCDS}$	°	449	61.89	15.38
Forward hood height at centerline	$h_{3,PCDS}$	cm	450	76.11	17.02
Forward hood height at centerline (wrap)	$w_{1,PCDS}$	cm	450	80.33	14.39
Hood length	HL_{PCDS}	cm	450	102.7	19.5
Transition point height at contact	$h_{4,PCDS}$	cm	450	85.90	15.56
Rear hood distance from ground at centerline (wrap)	$w_{2,PCDS}$	cm	450	184.79	22.22
Windshield base distance from ground at centerline (wrap)	$w_{3,PCDS}$	cm	450	194.49	22.58
Windshield top distance from ground at centerline (wrap)	$w_{4,PCDS}$	cm	450	271.82	21.08
Pedestrian physiology					
Age	y_{ped}	a	449	35.92	22.22
Body weight	$m_{ped,PCDS}$	kg	450	66.60	22.63
Body height	$h_{ped,PCDS}$	cm	450	162.79	19.48
Body mass index	BMI	kg/m^2	450	24.39	5.71
Knee height	$h_{k,ped,PCDS}$	cm	449	46.74	5.94
Hip height	$h_{h,ped,PCDS}$	cm	449	88.21	11.08
Shoulder height	$h_{s,ped,PCDS}$	cm	449	134.24	16.42
Knee height to bumper bottom height	$r_{1,PCDS}$	–	449	1.25	0.32
Knee height to bumper top height	$r_{2,PCDS}$	–	449	0.86	0.14
Hip height to transition point height	$r_{3,PCDS}$	–	449	1.06	0.23
Hip height to forward hood height	$r_{4,PCDS}$	–	449	1.21	0.27
Hip height to forward hood height (wrap)	$r_{5,PCDS}$	–	449	1.13	0.24

(continued)

Table A.2 (continued)

Variable name	Symbol	Unit	N	Mean	SD
Shoulder height to front hood height (wrap)	$r_{6,PCDS}$	–	449	1.72	0.37
Shoulder height to rear hood height (wrap)	$r_{7,PCDS}$	–	449	0.74	0.13
Height to transition point height	$r_{8,PCDS}$	–	450	1.95	0.42
Height to hood length	$r_{9,PCDS}$	–	450	1.67	0.54
Height to forward hood height (wrap)	$r_{10,PCDS}$	–	450	2.09	0.44
Height to rear hood height (wrap)	$r_{11,PCDS}$	–	450	0.89	0.16
Height to windshield base (wrap)	$r_{12,PCDS}$	–	450	0.85	0.15
Height to windshield top (wrap)	$r_{13,PCDS}$	–	450	0.60	0.09

Case numbers (N), mean, and standard deviation (SD) are given for the full data set

Table A.3 Recoding of non-continuous variables (GIDAS)

Original variable	N	Original coding	New variable	Symbol	N	New coding
Vehicle characteristics						
Passenger car, not further specified (n. f. s.)	22	0	Wedge-shaped (1, 2)		77	1
Wedge-shape 1	37	1	Not wedge-shaped (others)		838	0
Wedge-shape 2	40	2	**Wedge-shaped**	$type_{1,veh,GIDAS}$	**915**	
Pontoon shape 3	155	3				
Pontoon shape 4	547	4	Pontoon-shaped (3, 4, 5)		713	1
Pontoon shape 5	11	5	Not pontoon-shaped (others)		202	0
Box shape, n. f. s.	1	6	**Pontoon-shaped**	$type_{2,veh,GIDAS}$	**915**	
Box shape A	4	7				
Other	5	8	Van (6, 7, 10, 11)		47	1
Unknown	20	9	No van (others)		868	0
Box shape B	5	10	**Van**	$type_{3,veh,GIDAS}$	**915**	
Box shape C	37	11				
Two-wheeler	0	12				
Side of passenger car, brushed	8	13				
Side of passenger car, hit	18	14				
Side of truck, brushed	1	15				
Side of truck, hit	0	16				

(continued)

Table A.3 (continued)

Original variable	N	Original coding	New variable	Symbol	N	New coding
Passenger car rear end, n. f. s.	0	17				
Passenger car hatchback	0	18				
Passenger car fastback	3	19				
Passenger car notchback	1	20				
Truck rear end	0	21				
Between towing vehicle and trailer	0	22				
Design of opponent	**915**					
Pedestrian movement						
Yes, no further details	11	1	Low speed (1, 2, 3, 4)	$v_{ped,GIDAS}$	466	0
None	32	2	High speed (others)		449	1
Walked, no further details	267	3	**Walking: speed**		**915**	
Walked slowly	156	4				
Walked briskly	230	5	Low hazard (others)	$hazard_{ped,GIDAS}$	485	0
Ran	168	6	High hazard (2, 5, 6)		430	1
Other	6	8	**Walking: hazard**		**915**	
Unknown	45	9				

(continued)

Table A.3 (continued)

Original variable	N	Original coding	New variable	Symbol	N	New coding
Pedestrian collision speed	**915**					
Fallen, recumbent	6	0	Not fallen (others)		909	0
1 o'clock	18	1	Fallen (0)		6	1
2 o'clock	30	2	**Impact point (ped.): fallen**	$X_{ped,1}$	**915**	
3 o'clock	296	3				
4 o'clock	20	4	Not front (others)		802	0
5 o'clock	8	5	Front (1, 11, 12)		113	1
6 o'clock	16	6	**Impact point (ped.): front**	$X_{ped,2}$	**915**	
7 o'clock	5	7				
8 o'clock	14	8	Not back (others)		886	0
9 o'clock	346	9	Back (5, 6, 7)		29	1
10 o'clock	33	10	**Impact point (ped.): back**	$X_{ped,3}$	**915**	
11 o'clock	26	11				
12 o'clock	69	12	Not side (others)		176	0
unknown	28	99	Side (2, 3, 4, 8, 9, 10)		739	1
Pedestrian impact point	**915**		**Impact point (ped.): side**	$X_{ped,4}$	**915**	
Pedestrian physiology						
Male	452	3	Female (4, 5)		457	0
Female	455	4	Male (3)		452	1
Pregnant	2	5	**Sex**	G_{ped}	**909**	
Unknown	6	9				
Sex	**915**					

Table A.4 Recoding of non-continuous variables (PCDS)

Original variable	N	Original coding	New variable	Symbol	N	New coding
Vehicle characteristics (static)						
Convertible	5	1	Passenger vehicle (1–6)		241	1
2-door sedan, hardtop, coupe	50	2	No passenger vehicle (others)		128	0
3-door/2-door hatchback	32	3	**Body type of the vehicle**	$type_{veh,PCDS}$	**369**	
4-door sedan, hardtop	141	4				
5-door/4-door hatchback	8	5				
Station wagon	5	6				
Compact utility	20	14				
Large utility	2	15				
Utility station wagon	4	16				
Minivan	40	20				
Large van	14	21				
Step van or walk-in van	1	22				
Compact pickup	17	30				
Large pickup	30	31				
Unknown body type	0	99				
Body type	**369**					
Driver maneuvers and attention						
No avoidance actions	143	1	Braking (2–4, 8, 9, 99)		150	1
Braking (no lockup)	101	2	Not braking (others)		249	0
Braking (lockup)	64	3	**Avoidance: Braking**	$b_{driver,1}$	**369**	
Braking (lockup unknown)	2	4				

(continued)

Table A.4 (continued)

Original variable	N	Original coding	New variable	Symbol	N	New coding
Releasing brakes	0	5	Braking with lockup (3, 4)		64	1
Steering left	4	6	Not braking with lockup (others)		305	0
Steering right	2	7	**Avoidance: Braking with lockup**	$b_{driver,2}$	**369**	
Braking and steering left	28	8				
Braking and steering right	19	9	Releasing brakes (5)		0	1
Accelerating	1	10	Not releasing brakes (others)		369	0
Accelerating and steering left	0	11	**Avoidance: Releasing brakes**	$b_{driver,3}$	**369**	
Accelerating and steering right	0	12				
Other action	0	98	Accelerating (10–12)		1	1
Unknown	5	99	Not accelerating (others)		368	0
Attempted avoidance maneuver of the car	**369**		**Avoidance: Accelerating**	a_{driver}	**369**	
			Steering left (6, 8, 11)		32	1
			Not steering left (others)		337	0
			Avoidance: Steering left	$\delta_{driver,l}$	**369**	
			Steering right (7, 9, 12)		21	1
			Not steering right (others)		348	0
			Avoidance: Steering right	$\delta_{driver,r}$	**369**	
			Steering (6–9, 11, 12)		53	1
			Not steering (others)		316	0
			Avoidance: Steering	δ_{driver}	**369**	

(continued)

Table A.4 (continued)

Original variable	N	Original coding	New variable	Symbol	N	New coding
No driver present	0	0	Low complexity (0–4)		242	0
Going straight	229	1	High complexity (others)		127	1
Slowing or stopping in traffic lane	8	2	**Pre-event movement car: complexity**	c_{driver}	**369**	
Starting in traffic lane	3	3				
Stopped in traffic lane	2	4				
Passing or overtaking another vehicle	7	5				
Disabled or parked in travel lane	0	6				
Leaving a parking position	0	7				
Entering a parking position	0	8				
Turning right	32	9				
Turning left	74	10				
Making U-turn	1	11				
Backing up (other than for parking position)	0	12				
Negotiating a curve	3	13				
Changing lanes	5	14				
Merging	1	15				
Successful avoidance maneuver to a previous critical event	2	16				
Other	1	97				
Unknown	1	99				

(continued)

Table A.4 (continued)

Original variable	N	Original coding	New variable	Symbol	N	New coding
Pre-event movement of the car	**369**					
Full attention to driving	304	1	Driver not distracted (1)		304	0
Distracted by other occupant	6	2	Driver distracted (others)		65	1
Distracted by moving object in vehicle	1	3	**Driver's attention**	att_{driver}	**369**	
Distracted by outside person, object or event	21	4				
Talking on cellular phone/CB radio	2	5				
Sleeping or dozing while driving	3	6				
Other	21	8				
Unknown	11	9				
Driver attention	**369**					
Pedestrian physiology (incl. ratios)						
Male	186	1	Male (1)		186	1
Female—not reported pregnant	180	2	Female (2–6)		183	2
Female—pregnant—1st trimester	2	3	**Sex**	G_{ped}	**369**	
Female—pregnant—2nd trimester	1	4				

(continued)

Table A.4 (continued)

Original variable	N	Original coding	New variable	Symbol	N	New coding
Female—pregnant—3rd trimester	0	5	Not pregnant		366	0
Female—pregnant—term unknown	0	6	1st trimester (3)		2	1
Unknown	0	9	2nd trimester (4)		1	2
Sex	**369**		**Pregnancy**	P_{ped}	**369**	
Pedestrian movement						
Stopped	10	0	Walking away from lane (6)		2	1
Crossing road—straight	272	1	Not walking away from lane (others)		367	0
Crossing road—diagonally	58	2	**Direction ped.: away from lane**	$\omega_{ped,1}$	**369**	
Moving in road—with traffic	10	3				
Moving in road—against traffic	2	4	With; against traffic (3, 4, 7, 9)		15	1
Off road—approaching road	0	5	Not with; against traffic (others)		354	0
Off road—going away from road	2	6	**Direction ped.: with; against traffic**	$\omega_{ped,2}$	**369**	
Off road—moving parallel	3	7				
Off road—crossing driveway	9	8				
Off road—moving along driveway	0	9	Towards lane; crossing (1, 2, 5, 8)		339	1

(continued)

Table A.4 (continued)

Original variable	N	Original coding	New variable	Symbol	N	New coding
Other	1	98	Not towards lane; crossing (others)	$\omega_{ped.3}$	30	0
Unknown	2	99	**Direction ped.: towards lane; crossing**		369	1
Action of the pedestrian	369					
Not moving	12	0	Low speed (2, 3)		197	0
Walking slowly	178	1	High speed (others)		172	1
Walking rapidly	36	2	**Walking: speed (2, 3)**	$v_{ped,PCDS}$	369	
Running or jogging	136	3				
Hopping	0	4	Low hazard (others)		182	0
Skipping	0	5	High hazard (0, 2, 3, 7)		187	1
Jumping	0	6	**Walking: hazard**	$hazard_{ped,PCDS}$	369	
Falling/stumbling or rising	3	7				
Other	1	8				
Unknown	3	9				
Motion of the pedestrian	369					

(Several descriptions of the original variables have been shortened compared to the codebook [1])

Reference

1. UMTRI. (1994–1998). NASS Pedestrian Crash Data Study (PCDS) Codebook. Version 03Mar01. UMTRI Transportation Data Center, 2005.